applied mathematics practices
for the 21st century

FROM PERCENTAGES TO ALGEBRA

Using Authentic Problem Contexts

Kenneth Chelst, Ph.D.

Thomas Edwards, Ph.D.

Deborah Ferry

Marianne Srock

Special Acknowledgements:
Format and editing: Sheida Marashi
Design and layout: Susan Marie Dials

Applied Mathematics Practices for the 21st Century (AMP21) is an affiliation of professors and mathematics educators who share a common desire to bring relevance to K-12 mathematics by using authentic problem contexts in teaching and developing mathematics concepts and skills. The lead faculty are Kenneth Chelst (College of Engineering) and Thomas Edwards (College of Education) who are both professors at Wayne State University in Detroit, MI.

AMP21 is a Non-Profit developer and provider of curriculum that is aligned with the eight Standards for Mathematical Practice in the Common Core State Standards. Our team has published two textbooks for high school 1) Algebra and Mathematical Modeling and 2) Probability and Mathematical Modeling. Our new curriculum development focuses on middle school topics central to proportional reasoning: percentages, rates, ratios and proportions. Everything is presented in authentic problem contexts that involve making decisions. AMP21 offers professional development workshops in conjunction with local universities to help teachers and schools develop the programs needed to enable students to succeed in the global economy. In addition, this curriculum can form a basis for project based learning in mathematics.

First Printing, 2017

ISBN-13:
978-1547194445

ISBN-10:
1547194448

Applied Mathematics Practices (AMP21)
In conjunction with Wayne State University
www.appliedmathpractices.com

Contact: Kenneth Chelst
4815 Fourth St. (Room 2017)
Detroit, MI 48201
kchelst@wayne.edu

TABLE OF CONTENTS

Introduction	VI
Activity 1: Choosing the Better Deal	1
Activity 2: Making the Grade — When 100% is Possible	21
Activity 3: Free Throw Percentages	47
Activity 4: Dropping Out of High School — When 0% is Best	69
Activity 5: Military Special Ops	87
Activity 6: Growing Lawn Service — Compound Percentages	109
Activity 7: McFadden Restaurant — Changing Revenue	139
Activity 8: Managing Store Space	163
Activity 9: Multiple Flavor Ice Cream Sales	181
Activity 10: Ordering Hoodies — Multiple Percentages	199
Activity 11: Compound Growth of App Users	225
Activity 12: Reducing the Number of Homeless Veterans	253
Activity 13: Grades — Weighted Averages	279
Activity 14: Open24 Slushy Sales	305
Activity 15: Constructing Congressional Districts	321

Dear Teacher,

As mathematics teachers we are all challenged to provide our students with complex problems to solve. Our wish for their experience in mathematics is that they understand that real world scenarios involving mathematics often take more than one minute or one statement to solve. Where do we find those materials? Mathematics is used every day in business, industry, marketing, etc. to make decisions, plan ahead, and predict outcomes. Many textbook examples do not engage students or help them see the value of mathematics.

This book is our attempt to provide you with authentic, meaningful, engaging scenarios that showcase the importance of mathematics in the world. As state and national assessments (PSAT/SAT) are beginning to test more complex mathematics problems through the contexts of science and social studies, we find ourselves looking for practice to develop those critical thinking skills necessary for achievement.

According to the College Board website, the revised PSAT/SAT Mathematics Test "focuses in-depth on three essential areas of math: Problem Solving and Data Analysis, Heart of Algebra, and Passport to Advanced Math. Problem Solving and Data Analysis is about being quantitatively literate. It includes using ratios, percentages, and proportional reasoning to solve problems in science, social science, and career contexts. The Heart of Algebra focuses on the mastery of linear equations and systems, which helps students develop key powers of abstraction. Throughout the PSAT/SAT, you'll be asked questions grounded in the real world, directly related to work performed in college and career." This book focuses on using percentages to solve problems in business, industry, marketing, government, sports, and management. The Mathematics section on the PSAT/SAT "features multi-step applications to solve problems in science, social science, career scenarios, and other real-life situations. The test sets up a scenario and asks several questions that give you the opportunity to dig in and model it mathematically."

The scenarios in this book provide students with similar experiences. Students are given an authentic, meaningful scenario, asked to answer questions, dig deeper and model the situation mathematically. Some activities involve interpreting/analyzing data from a graph or table, while others involve creating a graph or table with given data or data they have collected themselves. Activities engage students in making transitions from percentages to algebraic expressions to solving equations.

This book provides a teacher guide for each of the activities. Each lesson begins with a brief Number Talk suggestion that directly relates to mindful mental mathematics for that particular lesson. The guide utilizes the lesson-planning framework of Thinking Through a Lesson Protocol which was developed through the collaborative efforts (lead by Margaret Smith, Victoria Bill, and Elizabeth Hughes) of the mathematics team at the Institute for Learning and faculty and students in the School of Education at the University of Pittsburgh. The lessons are structured in the Launch, Explore, Summary format. Suggestions have been provided in some of the columns; however, the "Who - Selecting/Sequencing" column (in the middle) is intentionally left blank for the teacher to make notes during the Explore portion of the lesson. It is intended for the teacher to sequence which groups/strategies will be presented and in what order during the Summary (most important) portion of the lesson. Much as a skillful conductor leads an orchestra, mathematics teachers must lead their students to a deeper understanding of the mathematics. This is done through collaborative discussions involving analyzing, critiquing, justifying, and making the mathematical connections. Each activity is expected to take between one and two class periods depending on your students' prior knowledge and the length of the class periods. Estimated time suggestions are provided, but can be easily adjusted to fit yours and your students' needs.

Enjoy!

Kenneth Chelst, Thomas Edwards, Debbie Ferry, Marianne Srock

Introduction to Percentages

Percentages are the most common measure of performance in diverse settings. Percentages are so useful that the word is among the top 2,000 words used in the English language. Every child sees a percentage when he or she receives a grade on an exam. Every food label reports the percentage of each daily nutritional requirement. All high schools and colleges work to reduce their dropout percentage. Store sales are advertised as a percentage off of the price. In basketball the percentage of free throws made is a standard statistic. Every election poll reports the percentage of people who will vote for a specific candidate.

Percentages are linked to fractions and decimals. To determine the equal percent of a fraction, the fraction is simply multiplied by 100%. For example, one quarter is equal to 25%.

$$1/4 \times 100\% = 25\%$$

To determine the equal percent of a decimal, the decimal is multiplied by 100%. Thus, 0.25 is equal to 25%.

$$0.25 \times 100\% = 25\%$$

To find a percentage of a number, the number is multiplied by the equal decimal or fraction. For example, during flu season, 25% of the children might be absent on any day. In a class of 20 students, how many students would be absent? This number is found by

$$(1/4) \times 20 = 5$$

or

$$0.25 \times 20 = 5$$

Thus, 5 children would be absent in a class of 20 students.

It is useful to remember the relationship between some common fractions and percentages. These are summarized in Table 1. Since a dollar corresponds to 100 cents, there is a natural correspondence between percent and US coins. We believe that thinking about common US coins can help with this memorization. For example, ¼ corresponds to 0.25 and 25%. Similarly, the quarter coin represents one-fourth of a dollar. It is also called 25 cents. Three quarters is 75% of a dollar. In pennies it is 75 cents. Three-tenths is 0.30 or 30%. It corresponds to three dimes or 30 cents.

Some of the most confusing percentages and decimal equivalents involve values between 0 and 0.10. One one hundredth is 0.01. Students may forget the 0 that appears after the decimal place and before the one. In writing one percent, there is no need for a zero before the 1. Similarly, 0.05 corresponds to 5%.

<table>
<tr><th>Fraction</th><th>Decimal</th><th>Percent</th><td rowspan="10">The coin amount that corresponds to a percent or fraction of a dollar</td><th>Coins</th><th>Cents</th></tr>
<tr><td>1</td><td>1.00</td><td>100%</td><td>1 dollar</td><td>100</td></tr>
<tr><td>3/4</td><td>0.75</td><td>75%</td><td>Three quarters</td><td>75</td></tr>
<tr><td>½ or 2/4</td><td>0.50</td><td>50%</td><td>Half dollar Two quarters</td><td>50</td></tr>
<tr><td>3/10</td><td>0.30</td><td>30%</td><td>Three dimes</td><td>30</td></tr>
<tr><td>1/4</td><td>0.25</td><td>25%</td><td>One quarter</td><td>25</td></tr>
<tr><td>1/5 or 2/10</td><td>0.20</td><td>20%</td><td>Two dimes</td><td>20</td></tr>
<tr><td>1/10</td><td>0.10</td><td>10%</td><td>One dime</td><td>10</td></tr>
<tr><td>1/20</td><td>0.05</td><td>5%</td><td>One nickel</td><td>5</td></tr>
<tr><td>1/100</td><td>0.01</td><td>1%</td><td>One penny</td><td>1</td></tr>
</table>

Table 1: Percentages and US coins

There are many common fractions that do not match US coins. For example, one-third is equal to the repeating decimal 0.333. There is no US coin equal to one-third of a dollar. The equivalent percent is 33.3%. When using percentages, analysts usually round to the nearest tenth of a percent or just the nearest percent. Similarly, one-seventh equals the repeating decimal with seven places, 0.142857142857... Its percent equivalent is 14.3%. One-eighth is simply 0.125 and does not need a repeating decimal representation. It is 12.5%. Table 2 matches common fractions with the nearest percentage.

		Percent	
Fraction	Decimal	Nearest 0.1%	Nearest %
1/3	0.3333…	33.3%	33%
1/6	0.1666…	16.7%	17%
1/7	0.142857…	14.3%	14%
1/8	0.125	12.5%	13%
1/9	0.1111…	11.1%	11%

Table 2: Common fractions and percentages that do not match coins

Summary of Examples: Percentages Everywhere

There are a total of 15 examples. They are designed to explore different ways of working with percentages. All of the examples are embedded in scenarios that involve decisions.

1. **Choosing the Better Deal**: The first example starts with the most common textbook application of percentages, a price discount. The example also includes a fixed dollar discount. The student will compare the two and determine which is the better discount for different priced meals. This is facilitated with the introduction of an *algebraic expression* to be evaluated. The student then uses an *algebraic equation* to determine when the two discounts save the same amount of money.

2. **Making the Grade — When 100% is Possible**: This example uses grades on tests that are reported as percentages, this allows for comparison of grades with a different number of questions. In this example, the student has received a score for his first exam. He wants to figure out was score he needs on the next exam to earn a specific letter grade. An algebraic equation is used to determine the minimum grade required to achieve his desired goal.

3. **Free Throw Percentages**: This example discusses how a competing team might use the free throw percentages for the Detroit Pistons as guide as to whom to foul late in the game. The major part of the example presents fictional data on free throws that the student uses to calculate the player's free throw percentage.

4. **Dropping Out of High School– When 0% is Best:** In the previous examples a higher percentage was better. In this example the focus is on the percentage of students who drop out before completing high school. A smaller percentage is better. Students are asked to calculate percentages to determine which dropout prevention program is better.

5. **Military Special Ops:** In this example, two officers discuss the challenges of volunteers passing two rigorous weeks of different types of training for a special mission. The student will evaluate whether it is better to have the training with the higher failure rate first or last. This includes an economic analysis of the order of training. Students will use

algebra to determine how many volunteers are needed to enter training in order to meet the need for 36 soldiers who have passed both weeks of training.

6. **Growing Lawn Service – Compound Percentages**: Growing a business is one of the few applications in which percentages can exceed 100%. In this example students will evaluate two different marketing programs. One program increases the number of customers by a percentage. Multiple weeks of growth illustrate the concept of compound percentages. The other marketing program adds a fixed number of customers each week. Algebra is used to estimate the number of customers several weeks later as the two programs are compared.

7. **McFadden Restaurant – Changing Revenue:** This example discusses the monthly sales of a restaurant. The example addresses the misconception that if sales increase by 20% one month and then decrease by 20% the next month, the final number is the same as the starting number. It is not. This example also uses algebra to determine total revenue used to calculate the franchise fee.

8. **Managing Store Space**: This example is actually a collection of several small examples. In each case the decision maker is allocating a resource based on the percentage demand for different products. In several examples, the resource is space in the store. In another example, it is advertising dollars. This collection introduces the student to a commonly used demographic factor, the percentage of women and men shoppers. It also discusses the mix of customers who prefer organic fruits and vegetables and are willing to pay more for organic foods.

9. **Multiple Flavor Ice Cream Sales:** In previous examples there were only two possibilities. In this example the manager is deciding how much of each flavor of ice cream to stock. There are four flavors to choose from each with a different percentage of demand. This example involves an economic analysis of the alternative plans for stocking flavors. Students will need to work with numbers that do not always produce whole numbers of a product. Algebra is introduced to determine a break-even point for ordering a whole liter of ice cream even if all of it cannot be sold.

10. **Ordering Hoodies – Multiple Percentages:** This example continues the development of contexts with more than two percentages. The primary decision is how many items of each size should be ordered. The student is introduced to the concept of a pie chart that is used to show and compare percentages of sizes for men and women. One complication is that the calculations often result in non-whole numbers. However, only a whole number of items can be ordered. The problem also involves working with pairs of percentages, sizes and colors.

11. **Compound Growth of App Users:** In this example, two high schoolers develop a successful game app. The number of users grows by 25% each month. The example involves compounding percentages to determine the number of users several months from now. The example involves extensive economic analysis. The developers are selling advertising in order to raise money to pay for living on campus rather than commuting. This example includes a number of graphs that the student is asked to read and interpret.

12. **Reducing the Number of Homeless Veterans:** Homelessness is a national problem that is worked on by both federal and state agencies. This example presents two alternatives for reducing homelessness among veterans. One program reduces homelessness by 17% a year. The other helps a fixed number of veterans each year. The example explores the relative effectiveness of the two programs over a multi-year time period. Graphs and algebraic models are an important aspect of this example.

13. **Grades — Weighted Averages:** This is an extension of the earlier example involving grades. In this example the various components of the grades are not equally weighted. In addition, there will contexts with three elements to calculating the grade. An algebraic equation is used to determine the minimum percentage needed on the final element to bring the overall grade above a specific threshold.

14. **Open24 Slushy Sales:** This example presents data on the demand for different sized iced drinks. Calculating overall performance of the store involves taking a weighted sum of multiple sized drinks. The store manager is trying to decide between two different ad campaigns that can impact sales of these iced drinks.

15. **Constructing Congressional Districts** – Combining Percentages: This example explores how the design of a congressional district could affect the likelihood of a Republican or Democrat winning the congressional election. The region is made up of 12 geographic units with different percentages of Republican and Democratic voters. The state must group these 12 units into four congressional districts. The student will evaluate two different designs for the four districts. The student is then challenged to create a different plan that is more fair to both sides.

All 15 examples use percentages to make decisions in a meaningful context. In designing these examples, we strove to include the use of other important math skills in a natural way. Every example presents data in a table format. A number of the latter examples also include line graphs and pie charts. Seven of the examples introduce algebraic expressions and equations to determine a specific value of interest. At the end of all of the examples, we present a simple project idea for collecting data related to the context of the example.

Representations Used in Percentage Examples

0	Introduction	Concepts	Table	Algebra	Graph Chart	Project Idea
1	Choosing the Better Deal	Compare percent and fixed value of coupons	Y	Solve Equation	N	Y
2	Making the Grade	Best is 100% - weighted scores	Y	Solve Equation	N	Y
3	Free Throw Percentages	Best is only 90%	Y	N	N	Y
4	Dropping Out of High School	Best is 0 %	Y	N	N	Y
5	Military Special Ops	Order of percent change no impact	Y	Solve Equation	N	Y
6	Growing Lawn Service	Compare percent and fixed value	Y	Solve Equation	N	Y
7	McFadden Restaurant	Fallacy of equal + and – percentage change	Y	N	Y	Y
8	Managing Store Space	More than two percentages and non-whole number answers	Y	N	N	Y
9	Multiple Flavor Ice Cream	More than two percentages and non-whole number answers	Y	N	Y	Y
10	Ordering Hoodies	Multiply two percentages and non-whole number answers	Y	N	Y	Y
11	Growth of App Users	Compound percentages and financial analysis	Y	Set up Formula	Y	Y
12	Homeless Veterans	Compound percent vs fixed	Y	Set up Formula	Y	Y
13	Grades	Weighted Scores - 3 values	Y	Solve Equation	N	Y
14	Open24 Slushy Sales	Weighted sum of percentages	Y	N	Y	Y
15	Congressional Districts	Combine geographies and their percentages	Y	N	Y	Y

Activity 1: Choosing the Better Deal

Mathematical Goals

The student will use percentages to determine which coupon is better.

In part I the student will:

- Read an advertisement and interpret the information
- Perform operations with percentages
- Complete a table
- Work with percentages in a meaningful context familiar to students

In part II, appropriate for students who have been introduced to basic algebra, the student will:

- Transition to the use of Algebra to answer a question
- Evaluate an algebraic expression for different values of the variable
- Solve an algebraic equation

Before the lesson (5-10 minutes)

Number talk possibilities:

Select two or three depending on student abilities.

- Find 10% of $30.00.
- Find 10% of $23.00.
- Find 10% of $23.50.
- Find 5% of $30.00.
- Find 5% of $23.00.
- Find 5% of $23.50.

Choosing the better deal

Jimmy's Coney Island sends out the coupon below in a weekly mailer.

Chris went to Coney for a quick lunch. His bill for the meal was $5.00. He handed the cashier the 15% off coupon.

1. How much money did he save? How much did he pay for lunch?

His friend Clarissa wondered why Chris had not used the $2 coupon instead. Chris told her to look carefully at the rule for using the $2 coupon.

2. Why couldn't Chris use the $2 coupon?

3. If Chris could have used the $2 coupon what percent of the bill would he have saved?

4. Why do you think Coney Island limited the use of the $2 coupon?

5. Ramon, Jennifer and Marianne were having a hearty breakfast at Jimmy's and the total bill came to \$35. Which coupon should they use to save the most money? Justify your answer by finding the savings for both coupons and then comparing them.

6. Donald was having breakfast alone and his total bill was \$10. Which coupon should he use to save the most money? Justify your answer by finding the savings for both coupons and then comparing them.

Chris' friend Clarissa came up with a different way to think about the problem. She wanted to organize her thinking. She let y represent the total bill for any meal. The letter y is called a *variable*, because its value can change. Different meals can have different total bills. To find 15% of a number, you can multiply the number by 0.15. She multiplies y by 0.15 to represent the amount taken off the total bill if they use the 15% off coupon:

$0.15y$ = the amount taken off the total bill if you use the 15% off coupon.

If a dinner costs \$30, the variable, y, is replaced by \$30 in the expression. Then the amount off is

0.15(\$30) = \$4.50, and the dinner costs \$30.00–\$4.50 = \$25.50

If a dinner costs \$18, then y is replaced in the expression by \$18. The expression representing the amount off is 0.15(\$18) = \$2.70, and the dinner costs \$18.00–\$2.70 = \$15.30.

7. Fill in Table 1 for y = \$11.00 and \$15.00.

Total Bill y	15% off Coupon		$2.00 off Coupon	
	Amount Off **$0.15y$**	**Final Bill** **$y - 0.15y$**	**Amount Off** **$2.00**	**Final Bill** **$y - 2.00$**
$5.00	(0.15)$5.00=$0.75	$5.00-$0.75=$4.25	Not Applicable	Not Applicable
$10.00	(0.15)$10.00=$1.50	$10.00–$1.50=$8.50	$2.00	$10.00–$2.00=$8.00
$11.00				
$15.00				
$18.00	(0.15)$18.00=$2.70	$18.00–$2.70=$15.30	$2.00	$18.00–$2.00=$16.00
$30.00	(0.15)$30.00=$4.50	$30.00–$4.50=$25.50	$2.00	$30.00–$2.00=$28.00
$35.00	(0.15)$35.00=$5.25	$35.00–$5.25=$29.75	$2.00	$35.00–$2.00=$33.00

Table 1: Comparison of two coupons for different priced meals

8. Try some different values for y in the table above to find the prices when the $2.00 off coupon is the better deal and when the 15% off coupon is the better deal. Is there any price when the two coupons result in the same deal? If so, what is that price?

Use the data in Table 1 to fill in the following two statements:

9. When the bill is less than _______ the better coupon is ______ off.

10. When the bill is more than _______ the better coupon is ______ off.

Part II – Algebra – Better than trial and error

By substituting numbers for the variable in the table, it is possible to see a pattern when one coupon is better than the other. However, this method does not easily find the exact value at which the two coupons are worth the same. In the above discussion, the representation $(0.15)y$ is called an algebraic expression. This algebraic expression is useful, because it represents the amount taken off any meal when the 15% off coupon is used. It also can be used to answer the question, "When is the amount taken off the same for both coupons?" To answer that question, we can write an *algebraic equation*. We set the 15% off algebraic expression equal to \$2, the constant amount taken off when using the other coupon.

$$0.15y = \$2$$

> *algebraic expression*: a mathematical expression that consists of variables, constant numbers and mathematical operations. (Addition and multiplication are examples of operations.) The value of an algebraic expression changes as the value of the variable changes.

We now need to find the value of y that makes the algebraic expression equal to \$2. Finding the correct value of y is what solving an algebraic equation means. To do this, we apply the same mathematical operations to both sides of the equation until we are left with just $1y$ on the left hand side. In the equation, y has been multiplied by 0.15. To undo multiplying by 0.15, we use the inverse operation, division. When we divide both sides of the equation by 0.15, the left hand side becomes 1y, because 0.15y/0.15 = 1y. The two sides of the equation are still equal, so

$1y = \$2 \div 0.15 \approx \$13.33.$

11. To show that this answer is correct calculate (0.15)(\$13.33). Is the answer \$2?

12. Which coupon would be better if the meal cost

  a. \$13.34?
  b. \$13.30?
  c. \$13.36?
  d. \$13.29?
  e. \$13.37?

Solving an algebraic equation. To solve an algebraic equation, change the equation so that the variable $1y$ is alone on one side of the equation, and the other side is a number. To do this, carry out the same mathematical operations (+, –, x, ÷) in the same order on both sides of the equation. Anytime you carry out the same operation on both sides of the equation, the quantities will still be equal. We can subtract or add the same number to each side of an equation, and the two sides will still be equal. We can multiply or divide both sides of an equation by the same number, and the two sides will still be equal. This process of performing the same operations on both sides of an equation to reach the goal of just 1y on one side and a number on the other is one part of learning algebra. When you learn the process, you have the ability to solve an equation. You will develop this skill as you study different kinds of algebraic equations. You will learn to recognize patterns to decide what set of operations should be applied to both sides of the equation.

13. Based on your previous answers, for which prices does
    a. the $2.00 off coupon give the most off?

    b. the 15% off coupon give the most off?

    c. the two coupons give the same amount off?

Jimmy's Coney Island determined the coupons they were offering reduced their profits too much. They decided to offer coupons of 10% off or $1.50 off.

14. For what priced meals should you use the 10% coupon? When should you use the $1.50 coupon? When are the savings for both coupons equal?

Project Idea:

Look for an advertisement that has a restriction on using the discount. What is the restriction? Does it make sense?

Look for advertisements that offer both a percent discount and a fixed dollar discount. Determine the range of prices for which the percent coupon is better than the fixed amount.

Practice problems

1. You are purchasing a gift online and the total bill is $24.52. The site offers a discount coupon of 20% off the total bill or free shipping and handling for orders over $20. The shipping and handling fee for this order is quoted at $5.29. Which offer saves you the most?

2. The online site above offers gift wrapping for $2.00. You decide to have your gift wrapped which brings the total bill to $26.52. Which offer saves you the most?

3. Rosie's Cafe offers a percentage discount on your birthday equal to your age. You must show valid ID with birthdate to qualify. You plan to eat at Rosie's on your 18th birthday. You also found a $4 off coupon for meals at Rosie's over $16.00. Your total bill is $17.65. Which discount saves you the most?

4. On your 18th birthday at Rosie's, what price for the total bill is the discount with the $4 off coupon and the 18% birthday discount the same?

5. A car dealership offers two choices as discounts. The first choice is $2,016 dollars off the price for 2016 model year cars. The second choice is an 11% discount on your vehicle of choice. You have decided to purchase a slightly used Kia from 2016. The purchase price is $18,500. Which discount offer saves you more money?

6. You are looking into buying a new car. You have a family and friends discount from Ford Motor Company and that will save you 17% off the sticker price. Ford is presently running a sale to reduce inventory. The sale involves a "cash discount" of $4,000. The vehicle you have in mind has a sticker price of $22,994. Which discount will save you more?

Activity 1: Teachers' guide

Thinking through a lesson protocol

Standards:

6.RP.A.3.C: Find a percent of a quantity as a rate per 100 (e.g., 30% of a quantity means 30/100 times the quantity); solve problems involving finding the whole, given a part and the percent.

6.EE.A.2.A: Write expressions that record operations with numbers and with letters standing for numbers. *For example, express the calculation "Subtract y from 5" as 5 - y.*

6.EE.A.2.C: Evaluate expressions at specific values of their variables. Include expressions that arise from formulas used in real-world problems. Perform arithmetic operations, including those involving whole-number exponents, in the conventional order when there are no parentheses to specify a particular order (Order of Operations).

6.EE.A.6: Use variables to represent numbers and write expressions when solving a real-world or mathematical problem; understand that a variable can represent an unknown number, or, depending on the purpose at hand, any number in a specified set.

6.EE.A.7: Solve real-world and mathematical problems by writing and solving equations of the form $x + p = q$ and $px = q$ for cases in which p, q and x are all nonnegative rational numbers.

Mathematical Practices:

MP1: Make sense of problems and persevere in solving them.
MP2: Reason abstractly and quantitatively.
MP3: Construct viable arguments and critique the reasoning of others.
MP6: Attend to precision.

Setting up the problem - Launch	
Selecting tasks/goal setting	(15-20 minutes) Briefly discuss or ask students in a whole group setting for ideas when percentages are used in real life. Prior to working this scenario, students should read the "Introduction to Percentages."
Questions	Can you think of a time that percentages are used? Think on your own, what is the relationship between 25% and a quarter? Share it with a friend. Students could build a table similar to the one given in the "Introduction to Percentages."

Monitoring student work - Explore		
Strategies and misconceptions-Anticipating	**Who - Selecting and sequencing**	**Questions and statements - Monitoring**
(20 minutes) Have students look at coupon from Jimmy's Coney Island and answer questions #1-6. Share results with whole group.		Explain the differences and the similarities between the coupons in the flyer. Note: Be sure to mention that the $2.00 off coupon may only be used when the total bill is $10 or more.
(20 minutes) Read text section after question #6 and before question #7 as a whole class. Then solve and discuss questions #6-10.		
(20 minutes) Read text section in Part II as a whole class. Then solve questions #11-14.		Depending on students' prior knowledge of algebraic expressions the time spent on this section may vary. Spend more time if this is their first exposure, less if they are comfortable with expressions.
		When solving equations, stress the importance of inverse operations (undoing) and performing the same operation on both sides of the equation sign to keep the quantities equal.

Monitoring individual student work - Explore		
Strategies and misconceptions-Anticipating	**Who - Selecting and sequencing**	**Questions and statements - Monitoring**
For off-task students or for students that seem to be self-conscious about you listening to them share.		I am just listening or looking to find out how you are working on the problem. This helps me think about what we will do later. What do you think is the Big Idea in the Introduction to Percentages reading?
For students that appear to be stuck.		Can you tell me a little about your reading? How could you describe the relationship between percents and coins/dollars? How would you describe the problem in your own words? What facts do you have? Could you try it with simpler numbers? Fewer numbers?
For students that want to ask you questions, these are ways to uncover their thinking and judge to what extent you want to respond.		Tell me what you've thought about so far. What do you know? Why are you interested in more information about that? Let me say a little about that part.

Managing the discussion – Summarize	
Parts of discussion - Connecting	**Questions and statements - Connecting**
Launching the discussion: Select the problems in questions #7-10 that students are struggling with or you wish to share out.	Will team 1 start us off by sharing one way of working on this problem? Please raise your hand when you are ready to share your solution. What did you do first when you were working on this problem? Let's start by clearing up a few things about the problem. Let's list some key parts in this problem? What was unclear in the problem?
Eliciting and uncovering student strategies	Joe would you be willing to start us off? What have you found so far? Can you repeat that? Can you explain how you got that answer? How do you know? Walk us through your steps. Where did you begin? Can you show us?
Focusing on mathematical ideas	Can you explain why this is true? Does this method always work? How is Bob's method similar to Kelly's method? What do all the solutions have in common? What would happen if I changed the numbers to _____?
Encouraging interactions	Do you agree or disagree with Kahlil's idea? What do others think? Would someone be willing to repeat what Tom just said? Would anyone be willing to add on to what Sue just said?
Concluding the discussion	Can anyone tell me some of the big ideas that we learned today? How would you explain what we learned today to a 5th grader? Some of the key points from our discussion today are . . . Tomorrow we will continue our exploration of ______ beginning with the idea from today that ___________.
Post lesson notes	You may wish to assign the practice problems that you feel would benefit the students.

Solutions to text questions

1. How much money did he save? How much did he pay for lunch?

   Saved $0.15 \times \$5 = \$0.75$*. Total cost* $= \$5.00 - \$0.75 = \$4.25$

2. Why couldn't Chris use the $2 coupon?

   The $2 coupon can only be used for purchases of $10 or more.

3. If Chris could have used the $2 coupon what percent of bill would he have saved?

   ($2/$5) (100%) = 40% savings

4. Why do you think Coney Island limited the use of the $2 coupon?

   The restaurant would make no money on the meal when the percent discount is large compared to the bill. Only when the bill is $10 or more do they feel there is enough profit to afford a $2 discount.

5. Ramon, Jennifer and Marianne were having a hearty breakfast at Jimmy's and the total bill came to $35. Which coupon should they use to save the most money? Justify your answer by finding the savings for both coupons and then comparing them.

   Solution 1

15% OFF Coupon:	*$2.00 OFF Coupon:*
You will save 15% of $35.00?	*You will save $2.00*
*0.15 * $35.00 = $5.25*	

   You save more with the 15% OFF Coupon

   Solution 2

15% OFF Coupon:	*$2.00 OFF Coupon:*
You will pay 85% of the bill.	*You will get $2.00 off your bill.*
$0.85 \times \$35.00 = \$29.75$	$\$35.00 - \$2.00 = \$33.00$

   The 15% OFF Coupon results in a smaller bill and saves the most money.

6. Donald was having breakfast alone and his total bill was $10. Which coupon should he use to save the most money? Justify your answer by finding the savings for both coupons and then comparing them.

   Solution 1

15% OFF Coupon:	*$2.00 OFF Coupon:*
You will save 15% of $10.00?	*You will save $2.00*
$0.15 \times \$10.00 = \$1.50$	

   The $2.00 OFF Coupon results in a smaller bill and saves the most money.

Solution 2

15% OFF Coupon: *\$2.00 OFF Coupon:*

You will pay 85% of the bill. *You will get \$2.00 off your bill.*

0.85 × \$10.00 = \$8.50 *\$10.00 - \$2.00 = \$8.00*

The \$2.00 OFF Coupon saves you the most money.

7. Fill in the Table 1 below for y = \$11.00 and \$15.00.

Total Bill y	***15% off Coupon***		***\$2.00 off Coupon***	
	Amount Off 0.15y	***Final Bill y – 0.15y***	***Amount Off \$2.00***	***Final Bill y – 2.00***
\$5.00	*0.15(\$5.00)=\$0.75*	*\$5.00–\$0.75 =\$4.25*	*Not Applicable*	*Not Applicable*
\$10.00	*0.15(\$10.00)=\$1.50*	*\$10.00–\$1.50 =\$8.50*	*\$2.00*	*\$10.00–\$2.00=\$8.00*
\$11.00	*0.15(\$11.00)=\$1.65*	*\$11.00–\$1.65 =\$9.35*	*\$2.00*	*\$11.00–\$2.00=\$9.00*
\$12.00	*0.15(\$12.00)=\$1.80*	*\$12.00-\$1.80 =\$10.20*	*\$2.00*	*\$12.00–\$2.00 =\$10.00*
\$13.00	*0.15(\$13.00)=\$1.95*	*\$13.00–\$1.95 =\$11.05*	*\$2.00*	*\$13.00–\$2.00 =\$11.00*
\$15.00	*0.15(\$15.00=\$2.25*	*\$15.00–\$2.25 =\$12.75*	*\$2.00*	*\$15.00–\$2.00 =\$13.00*
\$18.00	*0.15(\$18.00)=\$2.70*	*\$18.00–\$2.70 =\$15.30*	*\$2.00*	*\$18.00–\$2.00 =\$16.00*
\$30.00	*0.15(\$30.00)=\$4.50*	*\$30.00–\$4.50 =\$25.50*	*\$2.00*	*\$30.00–\$2.00 =\$28.00*
\$35.00	*0.15(\$35.00)=\$5.25*	*\$35.00–\$5.25 =\$29.75*	*\$2.00*	*\$35.00–\$2.00 =\$33.00*

Table 1: Comparison of two coupons for different priced meals

8. Try some different values for y in the table above to find the prices when the \$2.00 off coupon is the better deal and when the 15% off coupon is the better deal. Is there any price when the two coupons result in the same deal? If so, what is that price?

   Answers will vary. See the calculations added to the Table 1.

9. When the bill is less than _______ the better coupon is ______ off.

   When the bill is less than \$10 the better/only coupon is 15% off. For bills between \$10 and approximately \$13.29, the better coupon is the \$2.00 off coupon.

10. When the bill is more than _______ the better coupon is ______ off.

   When the bill is more than \$13.37, the better coupon is the 15% off.

Part II Algebra

11. To show that this answer is correct calculate (0.15)(\$13.33). Is the answer \$2?

   (0.15)(\$13.33) = \$1.9995 ≈ \$2.00; yes, when rounded-off to the nearest penny, which is the smallest coin denomination.

12. Which coupon would be better if the meal cost is

   a. \$13.34? *(0.15)(\$13.34) = \$2.001 ≈ \$2.00; the coupons are equal in value*

   b. \$13.30? *(0.15)(\$13.30) = \$1.995 ≈ \$2.00; the coupons are equal in value*

   c. \$13.36? *(0.15)(\$13.36) = \$2.004 ≈ \$2.00; the coupons are equal in value*

   d. \$13.29? *(0.15)(\$13.29) = \$1.9935 ≈ \$1.99; the \$2.00 off coupon is better*

   e. \$13.37? *(0.15)(\$13.37) = \$2.0055 ≈ \$2.01; the 15% off coupon is better*

13. Based on your answers to questions 5 and 6, for which prices does
   a. the \$2.00 off coupon give the most off? *Prices ≤ \$13.29*
   b. the 15% off coupon give the most off? *Prices ≥ \$13.37*
   c. the two coupons give the same amount off?
      Because of rounding to the nearest penny, the discount is the same for prices from \$13.30 to \$13.36

14. For what priced meals should you use the 10% coupon? When should you use the \$1.50 coupon? When are the savings for both coupons equal?

   To answer that question, we can write an algebraic equation. We set the 10% off algebraic expression equal to \$1.50, the constant amount taken off when using the other coupon.
   0.1y = \$1.50
   We divide both sides by 0.1.
   0.1y / 0.1= \$1.50/0.1
   y = \$15.00
   However, because of rounding, values close to \$15.00 will also yield a \$1.50 discount.

Solutions to practice problems

1. You are purchasing a gift online and the total bill is $24.52. The site offers a discount coupon of 20% off the total bill or free shipping and handling for orders over $20. The shipping and handling fee for this order is quoted at $5.29. Which offer saves you the most?

    20% of $24.52 is $4.90. That discount is less than $5.29, therefore, it would be better to take the discount for the shipping and handling of $5.29.

2. The online site above offers gift wrapping for $2.00. You decide to have your gift wrapped bringing the total bill to $26.52. Which offer saves you the most?

    20% of $26.52 is $5.30. That is one penny more than the shipping and handling discount of $5.29, therefore, to save the one penny, you would take the 20% off discount.

3. Rosie's Cafe offers a percentage discount on your birthday equal to your age. You must show valid ID with birthdate to qualify. You plan to eat at Rosie's on your 18th birthday. You also found a $4 off coupon for meals at Rosie's over $16.00. Your total bill is $17.65. Which discount saves you the most?

    The discount for the $17.65 bill would be 18% of $17.65 which is $3.177. The $4.00 off coupon would be more of a savings.

4. On your 18th birthday at Rosie's, what price for the total bill is the discount with the $4 off coupon and the 18% birthday discount the same?

    18% of "n" is $4.00 then 4.00/.18 = $22.22. In order for the 18% discount to match the $4.00 discount, the bill would have to be $22.22.

5. A car dealership offers two choices as discounts. The first choice is $2,016 dollars off the price for 2016 model year cars. The second choice is an 11% discount on your vehicle of choice. You have decided to purchase a slightly used Kia from 2016. The purchase price is $18,500. Which discount offer saves you more money?

    11% of $18,500 is $2,035. Comparing that to the $2,016 cash giveaway, the $2,035 saves you more, precisely $19 more.

6. You are looking into buying a new car. You have a family and friends discount for Ford Motor Company and that will save you 17% off the sticker price. Ford is presently running a sale to reduce inventory. The sale involves a "cash discount" of $4,000. The vehicle you have in mind has a sticker price of $22,994. Which discount will save you more?

    17% of $22,994 is $3,908.98. That is less than the $4,000 discount, so that would save you more, exactly $91.02 more.

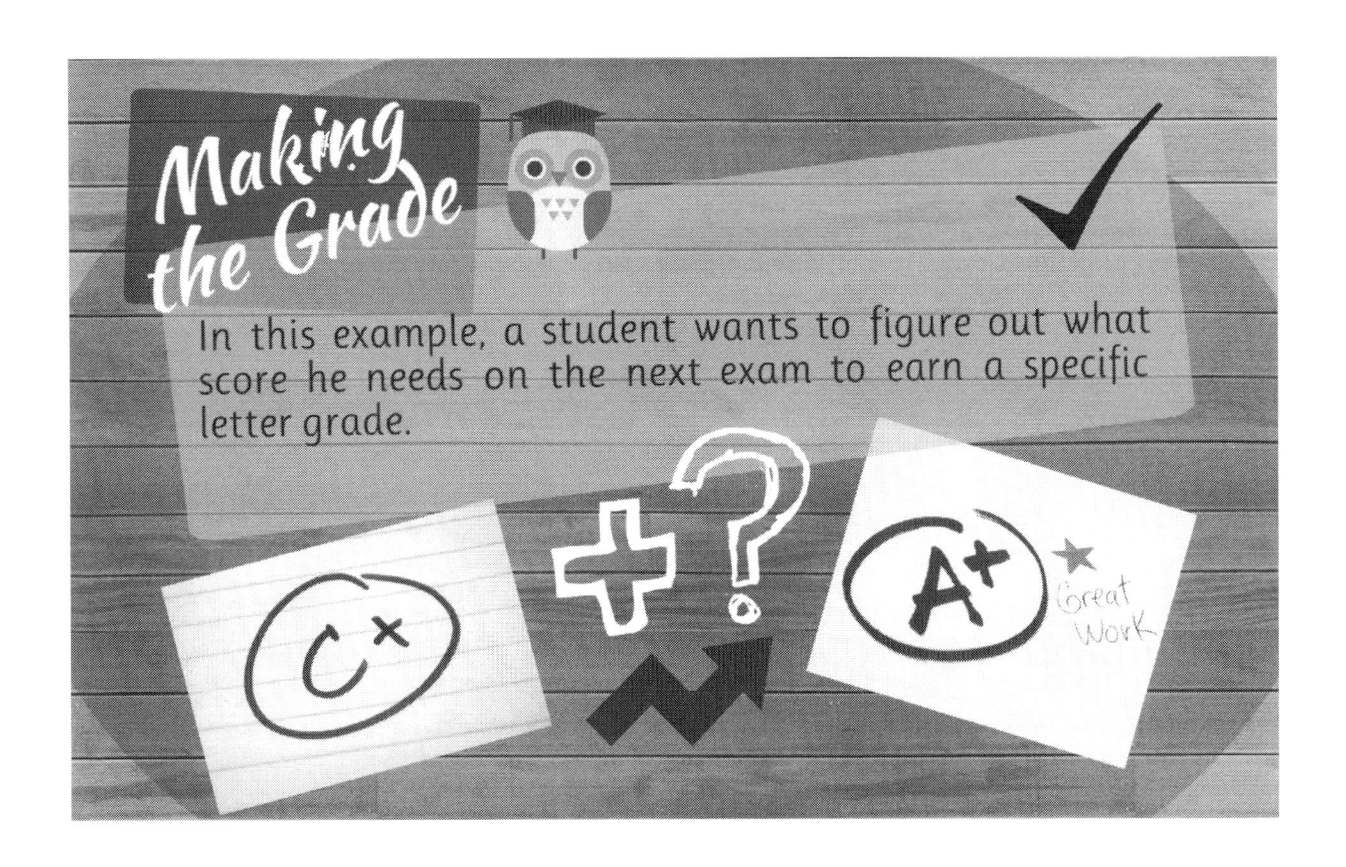

Activity 2:
Making the Grade
When 100% is Possible

Mathematical Goals

The student will use percentages to determine what it will take to earn course grades for various courses.

In part I the student will:

- Read a scenario and use data presented in table format
- Calculate different percentages
- Calculate averages of two numbers
- Evaluate an expression
- Work with percentages in a meaningful context familiar to students

In part II, appropriate for students who have been introduced to basic algebra, the student will:

- Transition into the use of Algebra to answer a question
- Solve an algebraic equation

Before the lesson (5-10 minutes)

Put the paper and pencil down and practice some mental mathematics.

Number talk possibilities:

Select two or three depending on student abilities.

- Change $\frac{3}{4}$ to a decimal, then to a percent.
- Change $\frac{3}{10}$ to a decimal, then to a percent.
- Change $\frac{2}{5}$ to a decimal, then to a percent.
- Change $\frac{1}{8}$ to a decimal, then to a percent.
- Change $\frac{1}{3}$ to a decimal, then to a percent.
- Change $\frac{2}{3}$ to a decimal, then to a percent.

Making the grade – when 100% is possible

Danielle got eight out of ten questions right on a science test. The fraction of questions she got right was $\frac{8}{10}$. To change this fraction to an equal decimal, we can divide the numerator by the denominator: $8 \div 10 = 0.8$. Now, to find an equal percent, we multiply the decimal by 100%:

$$0.8 \times 100\% = 80\%.$$

Why change a decimal to a percent? Decimals have no natural range of values. They go from negative infinity to positive infinity. Danielle might live 100 meters from school, or 2,000 meters, or anything in between. The number 1.00 has no special meaning. However, 100% does have a special meaning. Danielle told her father she finished 15 homework problems. Her father does not know if she just started, was in the middle, or finished the whole assignment. But, if she told him she did 100% of her homework, he would know everything was done. If she told her father she did anything less than 100% of her homework, he would know she still had some work to do.

Frank's busy week at school

This was a busy test week for Frank. He had tests in math, science, and history. (See Table 1) Each test had a different number of questions. The math test had the fewest questions, because each one took a lot of time. He answered 12 out of the 15 questions correctly. The fraction he got correct is $\frac{12}{15}$. Using his calculator, he found $\frac{12}{15} = 0.8$. Next, to multiply by 100%, he moved the decimal point two places to the right. His grade on the test was 80%.

	Math	Science	History
Correct Answers	12	20	19
Total Questions	15	40	20
Fraction Correct	$\frac{12}{15}$	$\frac{20}{40}$	$\frac{19}{20}$
Decimal	0.80	0.50	0.95
Percent	80%	50%	95%

Table 1: Frank's exam grades

The science test was all multiple choice questions. Frank answered many more questions right on the science test. He answered 20 questions correctly. However, there were 40 questions on the test. He only got half of the questions right. Thus, he earned 50% on this test. History is Frank's favorite subject. The test was a mixture of multiple choice questions and short answers. He got 19 out of the 20 questions right.

Percentages make it easy to compare tests with a different number of questions.

1. Which subject does Frank need to work hardest on to improve his grade?

2. In which subject is there not much room for improvement?

One hundred percent is a perfect score. However, it is not the only standard that is used. For example, in Frank's school, 65% is the passing grade on every test. Any student who fails a test must have a parent sign the test and return it to the teacher.

3. Will any of Frank's tests need a parent's signature? If so, which?

Randy's worries

Randy is not as studious as many of his classmates. His grades on the same tests are recorded in Table 2.

	Math	Science	History
Correct Answers	6	30	13
Total Questions	15	40	20

Table 2: Randy's exam grades

4. Will Randy need to have any of his tests signed by a parent? If so, which?

Randy thinks he might fail math. He did very poorly on the test. This test is one of two for math. The two tests will be averaged to determine his grade in math. He wonders how well he will have to do on the second test to get his average above 65%.

5. What fraction of the questions on the math test did Randy get right?

6. What decimal is equal to the fraction of the questions Randy got it right? What percent is that?

7. If he scores 75% on the second test, will his average be more than 65%?

8. If he scores 98% on the second test, will his average be more than 65%?

Randy asked his friend Alberta to help him find the lowest score he'd need to get his average up to 65% or higher. Alberta began by showing Randy how to determine the average for different scores on the second exam. She told him to use x to represent the score of the second test.

Step 1: Add the new score to the first score, which was 0.4, to find the total of the two scores

$0.4 + x$

Step 2: Calculate the average by dividing by 2.

$(0.4 + x) \div 2$

Randy used step 1 and step 2 to set up a table and evaluate different scores on the second exam.

9. Help Randy by filling in the missing entries in Table 3 below.

Second test score x	**Two scores added together** $0.40 + x$	**Average of two scores** $(0.40 + x) \div 2$
0.50	0.40 + .50=.90	.90 ÷ 2=.45
0.55	0.40 + 0.55 = 0.95	0.95 ÷ 2=.475
0.60		
0.65		
0.70		
0.75		

Table 3: Algebraic expressions calculate average

Part II – Algebra – Better than trial and error

By plugging in numbers, it is possible to see a pattern about the final grade. Randy soon got tired of the table. He still didn't know the exact answer. He asked Alberta if there was a shorter way to find the score he needed. In the above discussion, the representation $(0.40 + x) \div 2$ is called an algebraic expression. Alberta showed Randy how to use the algebraic expression to set up an equation to find the answer. Alberta said solving an equation would be shorter than his trial and error method.

> An *algebraic expression* is a mathematical expression that consists of variables, constant numbers and mathematical operations. (Addition and multiplication are examples of mathematical operations.) The value of an algebraic expression can change as the value of the variable changes.

Alberta used ×the expression for the average of the two tests, but she made one small change. Instead of using the ÷ sign, she wrote the division as a fraction. Then she made an equation by putting the expression for the average equal to the required average of 65%. After she changed 65% to 0.65, the equation looked like this:

$$\frac{0.40 + x}{2} = 0.65$$

To solve this equation, Alberta said, "First we multiply both sides of the equation by the same number, 2."

Randy said, "I see where the 2 came from, but why can we just multiply the left and right sides of the equation by 2?"

Alberta replied, "Imagine you had two equal amounts of money, like 2 quarters and 5 dimes. If you doubled both of them, wouldn't you still have two equal amounts of money?" Randy agreed that they'd still be equal, so he did the multiplication.

$$2\left(\frac{0.40+x}{2}\right) = 2(0.65)$$

This got rid of the fraction on the left side of the equation. The new equation was

$$0.40 + x = 1.30$$

To solve this new equation, they looked for the number, x, to add to 0.40 so that the left hand side of the equation will total 1.30. To figure that out, they subtracted 0.40 from both sides.

$$0.40 + x = 1.30$$

$$-0.40 = -0.40$$

$$x = 0.90$$

This left only x to the left of the equal sign and 0.90 on the right. This means that Randy will need 0.90, or 90% on the second test to get an average of 65%.

The second test is going to have 20 questions. Alberta showed Randy how to use another equation to find out how many questions he needs to get right. If n is the number of questions he gets right, then the fraction answered correctly is $\frac{n}{20}$. That fraction must equal 0.90.

$$\frac{n}{20} = 0.90$$

$$20\left(\frac{n}{20}\right) = 20(0.90)$$

$$n = 20 \times 0.90 = 18$$

Randy was worried about how much he would have to improve just to earn a passing grade. He was assigned a tutor to work with him every day for the next week. He did much better on the second test. He got 17 out of 20 questions right.

10. What grade did Randy earn for the class?

11. How could he use these test results to try to convince the teacher to give him a passing grade for the class?

What if the fraction is not easy to change into a percent?

In the examples above, all of the fractions could easily be changed to an equal decimal. Gerald got 10 out of the 15 questions on the math exam right, or two-thirds of the questions. The equal decimal is 0.6666…, and the 6 repeats forever. This is 66.66…% with repeating 6s. In solving problems, we don't usually go beyond one decimal place. Thus, 2/3 would be rounded to 66.7%. Often there is no need to be that accurate. The value can be rounded to the nearest whole percent. Gerald's teacher recorded his grade as 67%.

Gerald was also assigned a tutor. He wanted to do more than just pass math. He hoped to get an 85% average in math.

12. How high would he have to score to earn an 85% average?

Gerald's tutor thinks a more reasonable goal is to improve his average to 80%. Recall the final mathematics exam will contain 20 questions.

13. How many questions will Gerald need to get right to raise his average to 80% or higher?

14. Why do you think Gerald's tutor did not believe that 85% was a reasonable goal?

Project Idea: There are many ways that grades are calculated. Find out from your teacher how your grade is calculated in your math class and what grading scale is used. Calculate your current grade in your math class. To monitor your progress throughout the year, plot your current grade every two weeks on a graph.

Practice problems

John has already received an 85% and a 75% on two of the three tests in his mathematics class. John's teacher uses the following grading scale.

Percentage	Grade
100 – 90	A
89 – 80	B
79 – 70	C
69 – 60	D

1. What is John's current average in the class?

2. If John gets 100% on the third exam, what grade will he receive in the class?

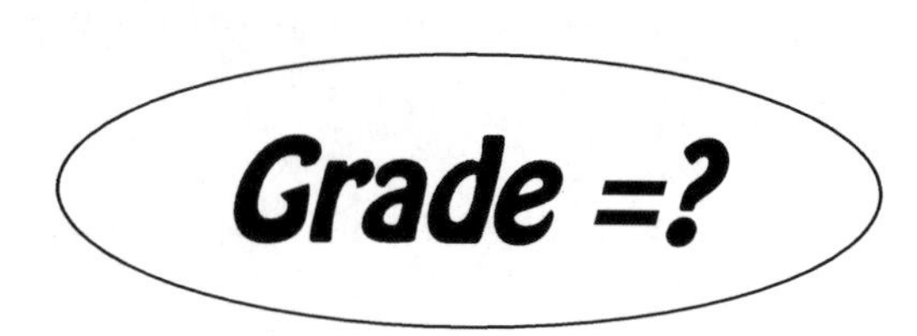

3. If John turns in a blank test and receives a zero on the third test, how will this affect his average?

4. Let x represent John's score on the third exam. Write a mathematical expression that can be used to calculate his average after receiving his score on the third exam.

5. John wants to be sure he earns at least a C grade in the class. Would he earn a C grade if he scored a 75 on the third exam? What if he scored a 65 on the third exam?

6. Write and solve an algebraic equation to determine the minimum percentage John needs on the third exam to receive a C in the class?

7. Determine the range of percentages John needs on the third test in order for him to receive a B in the class.

8. If the third test has a total of 20 questions, how many does John need to answer correctly to maintain his current average?

The local soccer organization plays 12 games in a season. Winning 60% or more of your games usually qualifies your team for the playoffs.

9. Your team, the Mustangs, has currently won 4 of the first 9 games. Is it possible for the Mustangs to make the playoffs? Why or why not?

10. Your friend's team, the Mavericks, have currently won 6 of the first 9 games. Is it possible for them to make the playoffs? Why or why not.

Activity 2: Teachers' guide

Thinking through a lesson protocol

Standards:

6.RP.A.3.C: Find a percent of a quantity as a rate per 100 (e.g., 30% of a quantity means 30/100 times the quantity); solve problems involving finding the whole, given a part and the percent.

7.RP.A.3: Use proportional relationships to solve multistep ratio and percent problems.

7.EE.B.4: Use variables to represent quantities in a real-world or mathematical problem, and construct simple equations and inequalities to solve problems by reasoning about the quantities.

Mathematical Practices:

MP1: Make sense of problems and persevere in solving them.

MP2: Reason abstractly and quantitatively.

MP3: Construct viable arguments and critique the reasoning of others.

Setting up the problem - Launch	
Selecting tasks/goal setting	(5 minutes) Briefly discuss or ask students in a whole group setting for ideas when percentages are used in place of a decimal or a fraction. Then have students read individually the first two paragraphs of the scenario, stopping at Frank's Busy Week at School. With a partner talk about what you think is the big idea in the reading.
Questions	Can you think of a time that percentages are used? When are they used in place of a decimal or a fraction? Why do you think they are used instead of a decimal or a fraction?

Monitoring student work - Explore		
Strategies and misconceptions- Anticipating	**Who - Selecting and sequencing**	**Questions and statements - Monitoring**
(10 minutes) Have students look at Table 1: Frank's exam grades and note two pieces of information found in the table. Have students share ideas in whole group setting while teacher records on board.		What are two pieces of information that you found in Table 1? (Teacher records on board).
Then read Frank's Busy Week at School and with a partner answer questions #1-3. Share out the results of questions #1-3.		As students share out their findings for questions 1-3 make sure that they justify their reasoning.
(30-40 minutes) In small groups have students continue to read Randy's Worries and answer questions #4-9. Monitor student work and check for understanding as the group's progress.		
(30-40 minutes) Read Part II - Algebra as a whole class. Carefully explain the Addition and Multiplication Properties of Equality while solving the equations. Have students solve questions #10-14 with a partner. Check for understanding while students are working. Share results as time allows.		Depending on students' prior knowledge of solving algebraic equations the time spent on this section may vary. Spend more time if this is their first exposure, less if they are comfortable with solving simple equations. Multiplying both sides of an equation by a number is an example of the Multiplication Property of Equality.

Monitoring individual student work - Explore		
Strategies and misconceptions- Anticipating	**Who - Selecting and sequencing**	**Questions and Statements - Monitoring**
For off-task students or for students that seem to be self-conscious about you listening to them share.		I am just listening or looking to find out how you are working on the problem. This helps me think about what we will do later. What do you think is the Big Idea in the Introduction to Percentages reading?
For students that appear to be stuck. Also for when you are having a difficult time understanding their strategies.		Can you tell me a little about your reading? How would you describe the problem in your own words? What facts do you have? Could you try it with simpler numbers?
For students that want to ask you questions, these are ways to uncover their thinking and judge to what extent you want to respond.		Tell me what you've thought about so far. What do you know? Why are you interested in more information about that? Let me say a little about that part.

Managing the discussion – Summarize	
Parts of discussion - Connecting	**Questions and statements - Connecting**
Launching the discussion: Select the problems in questions #10-14 that students are struggling with or you wish to share out.	Will team 1 start us off by sharing one way of working on this problem? Please raise your hand when you are ready to share your solution. What did you do first when you were working on this problem? Let's start by clearing up a few things about the problem. Let's list some key parts in this problem? What was unclear in the problem?
Eliciting and uncovering student strategies	Joe would you be willing to start us off? What have you found so far? Can you repeat that? Can you explain how you got that answer? How do you know? Walk us through your steps. Where did you begin? Can you show us?
Focusing on mathematical ideas	Can you explain why this is true? Does this method always work? How is Bob's method similar to Kelly's method? What do all the solutions have in common? What would happen if I changed the numbers to _____?
Encouraging interactions	Do you agree or disagree with Kahlil's idea? What do others think? Would someone be willing to repeat what Tom just said? Would anyone be willing to add on to what Sue just said?
Concluding the discussion	Can anyone tell me some of the big ideas that we learned today? How would you explain what we learned today to a 5th grader? Some of the key points from our discussion today are . . . Tomorrow we will continue our exploration of ______ beginning with the idea from today that ___________.
Post lesson notes	You may wish to assign the practice problems that you feel would benefit the students.

Solutions to text questions

1. Which subject does Frank need to work hardest on to improve his grade?

   Science: His first exam grade was the lowest, 50%

2. In which subject is there not much room for improvement?

   History: His first exam grade was 95% which is almost perfect.

3. Will any of Frank's tests need a parent's signature? If so, which?

   Science: His first exam grade was less than 65%

4. Will Randy need to have any of his tests signed by a parent? If so, which?

Math Grade $= 6/15 = 0.4 \rightarrow 40\%$	*Parent's signature required*
Science Grade $= 30/40 = 0.75 \rightarrow 75\%$	*No signature required*
History Grade $= 13/20 = 0.65 \rightarrow 65\%$	*No signature required*

5. What fraction of the questions on the math test did Randy get right?

   $6/15$

6. What decimal is equal to the fraction of the questions Randy got right? What percent is that?

   $6/15 = 0.4$

   $0.4 \times 100\% = 40\%$

7. If he scores 75% on the second test, will his average be more than 65%?

   No $(40\% + 75\%)/2 = 57.5\%$

8. If he scores 98% on the second test, will his average be more than 65%?

   Yes $(40\% + 98\%)/2 = 69\%$

9. Help Randy by filling in the missing entries in Table 3 below.

Second test score x	Two scores added together $0.40 + x$	Average of two scores $(0.40 + x) \div 2$
0.50	0.40 + 0.50=.90	0.90 ÷ 2 = .045
0.55	0.40 + 0.55 = 0.95	0.95 ÷ 2= 0.475
0.60	*0.40 + 0.60 = 1.00*	*1.00 ÷ 2 = 0.50*
0.65	*0.40 + 0.65 = 1.05*	*1.05 ÷ 2 = 0.525*
0.70	*0.40 + .70 = 1.10*	*1.10 ÷ 2 = 0.55*

Table 3: Algebraic expressions calculate average

10. What grade did Randy earn for the class?

17/20= 0.85 ➔ 85%

Average (40%+85%) / 2 = 62.5%

11. How could he use these test results to try to convince the teacher to give him a passing grade for the class?

- *One argument is that the two exams were not equally sized. If the two exams were combined the total number of questions is 35. He answered 6 questions correctly on the first exam and 17 on the second exam. In total he correctly answered 23 of 35 questions.*
  23 / 35 = 65.7% which is a passing grade
- *He could also argue the teacher should reward him for improving significantly from the first to the second exam.*

12. How high would he have to score to earn an 85% average?

$(0.67 + x) / 2 = 0.85$

$(0.67 + x) = 0.85(2) = 1.7$

$0.67 + x - 0.67 = 1.7 - 0.67$

$x = 1.03$ ➔ *103%*

13. Why do you think Gerald's tutor did not believe that 85% was a reasonable goal?

It is not possible to score more than 100% on an exam.

14. How many questions will Gerald need to get right to raise his average to 80% or higher? First calculate the percentage score that is needed.

$(0.67 + x)/2 = 0.8$

$(0.67 + x) = 0.8(2) = 1.6$

$0.67 + x - 0.67 = 1.6 - 0.67$

$x = 0.93 \rightarrow 93\%$

He will need to score at least 93%. The exam has 20 questions.

Number of questions correct $= 0.93 \times 20 = 18.6$

He will need to answer 18.6 questions correctly but there are no fractional questions. That means he will need to answer at least 19 questions correctly to achieve his goal of at least an 80% grade.

Solutions to practice problems

1. What is John's current average in the class?

   (85% + 75%) / 2 = 80%

   Students might reason that splitting the difference between 85% and 75% would be 80%.

2. If John gets 100% on the third exam, what grade will he receive in the class?

   (85% + 75% + 100%) / 3 = 86.666....%

   According to the grading scale, 80-89% is a B.

   John will receive a B in the class.

3. If John turns in a blank test and receives a zero on the third test, how will this affect his average?

   (75% + 85% + 0%) / 3

   160% / 3 = 53.333...%

   This will bring his current average of 80% down to a failing grade of 53.333...%.

4. Is it possible for John to get an A in the class? Why or why not?

   It is not possible for John to get an A in the class. Receiving a perfect score, 100% on the final test only earns him an 87% (if the teacher rounds up).

5. What is the minimum percentage John needs on the third exam to receive a C in the class?

   (85% + 75% + x%) / 3 = 70%

   (160% + x%)/3 = 70%

   160% + x% = 210%

   x% = 50%

6. Determine the range of percentages John needs on the third test in order for him to receive a B in the class.

To receive a B in the class, the minimum amount that John can receive on the third test is an 80%.

$(85\% + 75\% + x\%) / 3 = 80\%$

$(160\% + x\%) / 3 = 80\%$

$160\% + x\% = 240\%$

$x\% = 80\%$

$(85\% + 75\% + x\%) / 3 = 89\%$

$(160\% + x\%) / 3 = 89\%$

$160\% + x\% = 267\%$

$x\% = 107\%$

Since the maximum score is 100%, 107% is not possible. If you refer back to number 2, the maximum score of 100% was earned on the third test. The result was an average of 87%.

7. If the third test has a total of 20 questions, how many does John need to answer correctly to maintain his current average?

$x / 20 = 80\%$

$x = 80\%(20)$

$x = 0.80(20)$

$x = 16$

John will need to answer 16 out of the 20 questions correctly.

8. Your team, the Mustangs, has currently won 4 of the first 9 games. Is it possible for the Mustangs to make the playoffs? Why or why not?

   The Mustangs currently have 4 wins / 9 games or have won 44.444...% of the games.

   If the Mustangs continue to win, their record would improve to:

   5 wins / 10 games; 6 wins / 11 games; and finally 7 wins / 12 games.

   50%   55.545...%   58.333...%

   The Mustangs would typically NOT QUALIFY for the playoffs as the highest percent they could have would be 58%.

9. Your friend's team, the Mavericks, have currently won 6 of the first 9 games. Is it possible for them to make the playoffs? Why or why not.

   Scenario 1: If the Mavericks win their three remaining games, their record will be 9 wins / 12 games or a 75% record. They would QUALIFY for the playoffs.

   Scenario 2: If the Mavericks win two of their three remaining games, their record will be 8 wins / 12 games or a 66.666...% record. They would QUALIFY for the playoffs.

   Scenario 3: If the Mavericks win one of their three remaining games, their record will be 7 wins / 12 games or a 58.333...% record. They would NOT QUALIFY for the playoffs.

   Scenario 4: If the Mavericks don't win any of their three remaining games, their record will be 6 wins / 12 games or a 50% record. They would NOT QUALIFY for the playoffs.

Activity 3: Free Throw Percentages

Mathematical Goals

The student will use percentages to compare basketball players based on their ability to make free throw shots.

The student will use percentages to

- Read a scenario and use data presented in table format
- Perform operations on decimal numbers
- Change fractions to decimals to percentages
- Calculate different percentages
- Work with percentages in a meaningful context familiar to students

Before the lesson (5-10 minutes)

Put the paper and pencil down and practice some mental mathematics.

Number talk possibilities:

Select two or three depending on student abilities.

- 70 free throws made out of 100 attempts is ____%.
- 25 free throws made out of 50 attempts is ____%.
- 8 free throws made out of 10 attempts is _____%.
- 1 free throw made out of 5 attempts is ____%.
- 2 free throws made out of 8 attempts is ____%.

Free throw percentages

In most sports situations, 100% performance is unrealistic. In the 2015-2016 NBA season, the Golden State Warriors established a new record by winning 73 out of 82 games. They won 89% percent of their games. Similarly, no baseball team has ever won even 80% or more of their games in a season. In 1906 the Chicago Cubs won 116 out 152 games. Their winning percent was 76.3%.

Free throw percentages in the National Basketball Association (NBA)

Next, consider individual performances such as the percentage of free throws made. Only three NBA players out of thousands have managed to maintain above a 90% success rate through their careers. Poor free throw shooters are successful less than 60% of their attempts. Andre Drummond of the Pistons is ranked among the worst free throw shooters in the history of the NBA. His career percentage of success is less than 40%. Table 1 summarizes his free throw percentages for his four years in the NBA.

	Free throws		
Season	**Made**	**Attempted**	**Percentage**
2012-2013	59	159	37.1
2013-2014	137	328	41.8
2014-2015	142	365	38.9
2015-2016	208	586	35.5
Career	**546**	**1438**	**38.0**

Table 1: Andre Drummond's free throw percentage

Near the end of a game, some NBA teams try to take advantage of poor free throw shooters. They deliberately foul the poor shooter. They hope the shooter will make no more than one of his two free throws. After the free throws, the fouling team then has an opportunity to score two or even three points. This strategy was made famous when used against Shaquille O'Neal. Shaq's career free throw percentage success rate was 52.7%. This strategy was called the Hack-a-Shaq. Hack is slang for a personal foul.

1. In Table 1 what were the best and worst seasons for Andre Drummond?

2. In Table 1, describe a big change in the data for 2015-2016 compared to the previous two seasons.
3. What do you think caused the change?

Imagine that the Pistons have the five players listed in Table 2 on the court at the same time.

	Free throws		
Player	**Made**	**Attempted**	**Percentage**
Marcus Morris	203	271	74.9%
Kentavious Caldwell-Pope	185	228	81.1%
Andre Drummond	208	586	35.5%
Reggie Jackson	291	337	86.4%
Aron Baynes	126	165	76.4%

Table 2: Free throw percentages in 2015-2016 - five Detroit Pistons

4. How much worse is Andre Drummond than the Detroit Pistons' player with the next lowest free throw percentage?

5. Which player should the opposing team avoid fouling?

The strategy of simply fouling is boring for the fans. This was of special concern to the NBA towards the end of a close game. The NBA made a special rule for the final two minutes of the game. The goal was to prevent a team from deliberately fouling a player who does not even have the ball. If a player who does not have the ball is fouled, he first shoots two free throws. Then the ball is given back to the same team. The Pistons try to keep the ball away from Andre Drummond in the final two minutes.

6. Which Detroit Pistons should handle the ball most often in the final two minutes?
7. What rule changes would you propose to discourage strategies like the Hack-a-Shaq?

Daniella's free throw shooting

Daniella Kidman is an outstanding rebounder for the Duke Middle School girls' basketball team. However, she is one of the poorer free throw shooters on the team. The data in Table 3 records 30 pairs of shots Daniella took over the last ten games. An X means she made the free throw; an O means she missed the free throw.

	Shot				Shot				Shot		
Pair	**1**	**2**	**Points**	**Pair**	**1**	**2**	**Points**	**Pair**	**1**	**2**	**Points**
1	X	O	1	11	X	X	2	21	X	X	2
2	O	X	1	12	O	O	0	22	X	O	1
3	X	X	2	13	X	X	2	23	X	X	2
4	X	X	2	14	O	O	0	24	O	X	1
5	O	O	0	15	X	X	2	25	O	O	0
6	X	O	1	16	X	X	2	26	O	O	0
7	O	O	0	17	O	O	0	27	X	O	1
8	X	X	2	18	X	X	2	28	X	X	2
9	X	X	2	19	O	O	0	29	X	O	1
10	O	X	1	20	X	X	2	30	X	X	2

Table 3: Daniella's free throw shooting

8. What is Daniella's free throw percentage for the shots recorded in Table 3?

9. Would you recommend that opposing teams repeatedly foul Daniella? Why or why not?

Some players perform very differently on the first and second free throw.

10. Calculate Daniella's free throw percentage on the first shot?

11. Calculate the percentage on the second shot?

12. Is there a big difference in her percentage on the first and second shots?

The ideal outcome for a team that fouls Daniella is for her to miss both free throws.

13. What percentage of times did Daniella miss both free throws?

But the strategy does not work if Daniella makes both free throws.

14. What percentage of times did she make both free throws?

Project idea:

Set up stations in the room where students can shoot paper wads at waste baskets. Have them record the number they make out of 4, 5, 10, 12, and 25. Calculate and record percentages. As a class compare data and have students write a paragraph justifying which classmate should keep the ball at the end of the game (the best free throw shooter).

Practice problems

Elaine's Eagles Tennis Team

In tennis the person serving the ball has up to two chances to make the ball land within the required box of the opponent. In general, if the first serve does not land in the box, the player serves more slowly and carefully on the second attempt. Because the first serve is faster, the server has a better chance of winning the point when the first serve lands in the box. If the first serve is so fast that the opponent is not even able to hit it back, this is called an ace.

	First serve points won (Including aces)	Second serve points won	Total points won on serves	Percentage of total points won
Player 1	15 points/20 serves	4/9	19/29	65.5%
Player 2	10/18	5/12		
Player 3	6/18	8/11		
Player 4	14/21	9/10		

Table 4: Service points during a tournament

Use the data in the table above to answer the following questions about Elaine's Tennis Team.

1. During the Eagles last tennis tournament, the team statistician collected the data in the table above. Complete the table for the statistician.
2. Which player had the better percentage for Total Points on their serve?
3. Which player performed the best in the category of First Serve Points Won for this tournament? Explain how you determined your conclusion.
4. Which player performed the best in the category of Second Serve Points Won for this tournament? Explain how you determined your conclusion.

National Football Conference (NFC) team records

The data below was collected on October 28, 2016 from the Score app.

	Win-Loss-Tie	**Passing yards (average per game)**	**Rushing yards (average per game)**	**Points (average per game)**
Detroit Lions (NFC North)	4-3-0	273.0	111.4	24.3
Arizona Cardinals (NFC West)	3-3-1	253.9	125.4	15.7
Green Bay Packers (NFC North)	4-2-0	239.3	104.8	23.3
Seattle Seahawks (NFC West)	4-1-1	226.0	84.2	14.0

Table 5: Statistics for four NFC teams on October 28, 2016

5. Which team, on average, has the most total yards per game (rushing and passing)?
6. Find the percentage of passing yards to total yards for each team above.

7. Find the percentage of rushing yards to total yards for each team above.

8. Teams strive to have a balanced offense. The closer the passing and running percentages are the more balanced the offense. Which team had the most balanced offense? Which had the least balanced offense?

9. Find the percentage of wins to games played for each of the teams above.

10. Write two interesting facts from this data.

Activity 3: Teachers' guide

Thinking through a lesson protocol

Standards:

6.RP.A.3.C: Find a percent of a quantity as a rate per 100 (e.g., 30% of a quantity means 30/100 times the quantity); solve problems involving finding the whole, given a part and the percent.

7.RP.A.3: Use proportional relationships to solve multistep ratio and percent problems.

Mathematical Practices:

MP1: Make sense of problems and persevere in solving them.

MP2: Reason abstractly and quantitatively.

MP3: Construct viable arguments and critique the reasoning of others.

Setting up the problem - Launch	
Selecting tasks/goal setting	(5 minutes) Briefly discuss or ask students in a whole group setting for experiences they have had in shooting free throws and finding percentages. Other sports percentages could be included in this discussion. Then have students read individually the text stopping at Question #1. With a partner talk about what you think is the big idea in the reading and discuss what information is reported in Table 1.
Questions	How could we check to see if the percentages in Table 1 have been calculated correctly? Why do you think percentages are used instead of a decimal or a fraction?

Monitoring student work - Explore		
Strategies and misconceptions- Anticipating	**Who - Selecting and sequencing**	**Questions and statements - Monitoring**
(20 minutes) Have students answer questions # 1-7 on their own. Share out the results of questions #1-7. Students share out different strategies.		How can you justify each of your answers using information from the table? Teacher may wish to spend some time making sure that students understand the strategies explained.
Monitor student work and check for understanding as the group's progress. Have students read individually the Daniella free throw shooting section and review Table 3.		Discuss how information is recorded in Table 3. What information does the table provide?
Have students solve questions #8-13 in small groups. Share strategies used with whole group.		While sharing out be sure to watch for students that may have used a correct strategy for solving but misread or miscounted information from the table and therefore have a different answer than others.

Monitoring individual student work - Explore		
Strategies and misconceptions-Anticipating	**Who - Selecting and sequencing**	**Questions and statements - Monitoring**
For off-task students or for students that seem to be self-conscious about you listening to them share.		I am just listening or looking to find out how you are working on the problem. This helps me think about what we will do later. What do you think is the Big Idea in the reading?
For students that appear to be stuck. Also for when you are having a difficult time understanding their strategies.		Can you tell me a little about your reading? How would you describe the problem in your own words? What facts do you have? What information do you have in the table? Could you try it with simpler numbers? Fewer numbers?
For students that want to ask you questions, these are ways to uncover their thinking and judge to what extent you want to respond.		Tell me what you've thought about so far. What do you know? Why are you interested in more information about that? Let me say a little about that part.

Managing the discussion – Summarize	
Parts of discussion - Connecting	**Questions and statements - Connecting**
Launching the discussion: Select the problems in questions #8-13 that students are struggling with or you wish to share out.	Will team 1 start us off by sharing one way of working on this problem? Please raise your hand when you are ready to share your solution. What did you do first when you were working on this problem? Let's start by clearing up a few things about the problem. Let's list some key parts in this problem? What was unclear in the problem?
Eliciting and uncovering student strategies	Joe would you be willing to start us off? What have you found so far? Can you repeat that? Can you explain how you got that answer? How do you know? Walk us through your steps. Where did you begin? Can you show us?
Focusing on mathematical ideas	Can you explain why this is true? Does this method always work? How is Bob's method similar to Kelly's method? What do all the solutions have in common? What would happen if I changed the numbers to _____?
Encouraging interactions	Do you agree or disagree with Kahlil's idea? What do others think? Would someone be willing to repeat what Tom just said? Would anyone be willing to add on to what Sue just said?
Concluding the discussion	Can anyone tell me some of the big ideas that we learned today? How would you explain what we learned today to a 5th grader? Some of the key points from our discussion today are . . . Tomorrow we will continue our exploration of ______ beginning with the idea from today that __________.
Post lesson notes	You may wish to assign the practice problems that you feel would benefit the students.

Solutions to text questions

1. In Table 1 what were the best and worst seasons for Andre Drummond?

   Andre's best season was 2013-2014 with 41.8% of free throws made because 41.8% is the largest percentage. Andre's worst season was 2015-16 with only 35.5% of free throws made because of the four seasons 35.5% is the lowest percentage.

2. In Table 1, describe a big change in the data for 2015-2016 compared to the previous two seasons.

   In 2015-2016 Andre attempted many more free throws. He attempted over 200 more free throws than in the previous two seasons. He also made at least 50 more than in the previous two seasons, however his percentage is lower than either of the previous two seasons.

3. What do you think caused the change?

   Possible reasons: Teams became more aware of how poor a free throw shooter he was. He was fouled more often during the time he played in the games. It is possible Andre played in more games during the season or he was in each of the games for more minutes.

4. How much worse is Andre Drummond than the Detroit Pistons' player with the next lowest free throw percentage?

   The next lowest percentage is 74.9% by Marcus Morris. Andre is 39.4% worse than Marcus (74.9% - 35.5% = 39.4%).

5. Which player should the opposing team avoid fouling?

   Reggie Jackson because he has the highest percentage at 86.4%.

6. Which Detroit Pistons should handle the ball most often in the final two minutes?

   Reggie Jackson because he has the highest percentage for making free throws. In order from who should handle most to least it would be Reggie, Kentavious, Aron, Marcus and Andre because percentages from highest to lowest are 86.4%, 81.1%, 76.4%, 74.9%, and 35.5%.

7. What rule changes would you propose to discourage strategies like the Hack-a-Shaq?

   Anything is acceptable. Students *might suggest 3 or 4 free throws as opposed to only 2 in the last 2 minutes of the game. They might suggest no matter who is fouled, the team that is fouled gets to choose which player will shoot the free throws (either in game or on bench).*

8. What is Daniella's free throw percentage for the shots recorded in Table 3?

Since there are 30 pairs of shots recorded, there are 60 shots attempted. Counting the X's for made shots, there are 36 made shots. Another way would be to add the numbers in the 'Points' columns, also 36. So, 36 made out of 60 attempted is $\frac{36}{60}$ *as a fraction. I can divide 36 by 60 to get the decimal 0.6 and change it to a percent by multiplying by 100%. Daniella made 60% of her free throws.*

Another way to think about it is 36 out of 60 is equal to what out of 100. I could use the proportion $\frac{36}{60} = \frac{x}{100}$ *to solve.*

$60x = (36)(100)$

$60x = 3600$

$x = 60$

60 out of 100 is the same as 60%

I could also set up a ratio table to solve this.

Made	*36*	*6*	*60*	
Attempted	*60*	*10*	*100*	

Daniella made 60% of her free throws

9. Would you recommend that opposing teams repeatedly foul Daniella?

Although I am not a big fan of intentionally fouling a player, the paragraph above Table 3 indicated that Daniella is one of the poorer free throw shooters on the team so as far as the opposing team is concerned she would be a reasonable target to foul if she has the ball.

In the opening of the example it mentioned that poor free throw shooters are successful less than 60% of their attempts. Daniella is at exactly 60%, the next few attempts will indicate if she is over or under 60%.

10. Calculate Daniella's free throw percentage on the first shot?

There were 30 first shots given in the table. I counted the X's in column '1' only and got 19. So Daniella got 19 out of 30 attempts on first shots or $\frac{19}{30}$. *To change this fraction to a decimal I would divide 19 by 30 to get the repeating decimal 0.633333... To change to a percent multiply by 100% to get 63.3% rounded to the nearest tenth.*

11. Calculate Daniella's free throw percentage on the second shot?

There were 30 second shots given in the table. I counted the X's in column '2' only and got 17. So Daniella made 17 out of 30 attempts on second shots or $\frac{17}{30}$*. To change this fraction to a decimal I would divide 17 by 30 to get the repeating decimal 0.566666... To change to a percent multiply by 100% to get 56.7% rounded to the nearest tenth.*

12. Is there a big difference in her percentage of the first and second shots?

There is a 6.6% difference in first and second shots. This would be roughly 7 out of every 100 shots. I probably would not think of it as a big difference.

13. What percentage of times did Daniella miss both free throws?

I looked in the "Points" column of Table 3 and counted the number of zeros indicating both shots were missed. Of the 30 pairs, 8 times zero points were scored. So the fraction $\frac{8}{30}$ *changed to a percent is roughly 26.7%.*

14. What percentage of times did she make both free throws?

I looked in the "Points" column of Table 3 and counted the number of 2's indicating both shots were made. Of the 30 pairs, 14 times two points were scored. So the fraction $\frac{14}{30}$ *changed to a percent is roughly 46.7%.*

Solutions to practice problems

Elaine's Eagles Tennis Team

	First serve points won (including aces)	Second serve points won	Total points won on serves	Percentage of total points won
Player 1	15 points/20 serves	4/9	19/29	65.5%
Player 2	10/18	5/12	*15/30*	*50.0%*
Player 3	6/18	8/11	*14/29*	*48.3%*
Player 4	14/21	9/10	*23/31*	*74.2%*

Table 4: Service points during a tournament

Use the data in the table above to answer the following questions about Elaine's Tennis Team.

1. During the Eagles last tennis tournament, the team statistician collected the data in the table above. Complete the table for the statistician.

   See table above.

2. Which player had the better percentage of Total Points on their serve?

   Player 4 had 74.2% of points scored on their serve.

3. Which player performed the best in the category of First Serve Points Won for this tournament?
   Explain how you determined your conclusion.

   Player 1. Player 1 is 75%, Player 2 is 66.7%, Player 3 is 33.3%, and Player 4 is 66.7%.

4. Which player performed the best in the category of Second Serve Points Won for this tournament? Explain how you determined your conclusion.

   Player 4. Player 1 is 44.4%, Player 2 is 41.7%, Player 3 is 72.7%, and Player 4 is 90%.

National Football Conference (NFC) team records

The data below was collected on October 28, 2016 from the Score app.

	Win-Loss-Tie	Passing yards (average per game)	Rushing yards (average per game)	Points (average per game)
Detroit Lions (NFC North)	4-3-0	273.0	111.4	24.3
Arizona Cardinals (NFC West)	3-3-1	253.9	125.4	15.7
Green Bay Packers (NFC North)	4-2-0	239.3	104.8	23.3
Seattle Seahawks (NFC West)	4-1-1	226.0	84.2	14.0

Table 5: Statistics for four NFC teams on October 28, 2016

5. Which team on average, has the most total yards per game (rushing and passing)?

   Detroit 384.4, Arizona 379.3, Green Bay 344.1, Seattle 310.2. The team with the highest total yards per game on average (rushing and passing) is the Detroit Lions.

6. Find the percentage of passing yards to total yards for each team above.

   Detroit Lions 273.0/384.4 = 71.0%
   Arizona Cardinals 253.9/379.3 = 66.9%
   Green Bay Packers 239.3/344.1 = 69.5%
   Seattle Seahawks 226.0/310.2 = 72.9%

7. Find the percentage of rushing yards to total yards for each team above.

   Detroit Lions 111.4/384.4 = 29.0%
   Arizona Cardinals 125.4/379.3 = 33.1%
   Green Bay Packers 104.8/344.1 = 30.5%
   Seattle Seahawks 84.2/310.2 = 27.1%

8. Teams strive to have a balanced offense. The closer the passing and running percentages are the more balanced the offense. Which team had the most balanced offense? Which had the least balanced offense?

The Arizona Cardinals had the most balanced offense with passing accounting for only 66.9% of the total yards. The Seattle Seahawks had the least balanced offense. Passing accounted for 72.9% of the total yards.

9. Find the percentage of wins to games played for each of the teams above.

   Detroit Lions 4/7 or 57.1%
   Arizona Cardinals 3/7 or 42.9%
   Green Bay Packers 4/6 or 66.7%
   Seattle Seahawks 4/6 or 66.7%

10. Write two interesting facts from this data.

    Answers will vary.

    Examples: The Green Bay Packers and Seattle Seahawks have the same winning percentages. However, they don't have the same Win-Loss-Tie record. The Lions and Cardinals have played 7 games while the Packers and the Seahawks have only played 6 games.

Activity 4:
Dropping Out of High School
When 0% is Best

Mathematical Goals

The student will use unit rates, ratios and percentages to compare strategies for decreasing student dropout rates at two high schools.

The student will:

- Read a scenario and use data presented in table format
- Perform operations with ratios and rates
- Change fractions to decimals to percentages
- Calculate different percentages
- Transition into the use of Algebra to answer a question
- Work with percentages in a meaningful context familiar to students

Before the lesson (5-10 minutes)

Put the paper and pencil down and practice some mental mathematics.

Number talk possibilities:

Select two or three depending on student abilities.

- 16 out of 200 is ___ %.
- 9 out of 50 is ___ %.
- 25 out of 200 is ____ %.
- 13.5 out of 50 is ___ %.
- 45 out of 500 is ____%.

Dropping out – When 0% is best

The Greenwood School District has two high schools, Byron Day High and Brad Knight High. Each school has a problem of students dropping out in their senior year before graduating. In 2014-2015, 21.1% of the seniors at Byron Day dropped out. At Brad Knight 25% of the seniors dropped out that same year. The dropout data are in Table 1.

Byron Day High School			Brad Knight High School		
Started	**Dropped out**	**Dropout %**	**Started**	**Dropped out**	**Dropout %**
180	38	21.1%	180	45	25.0%

Table 1: 2014-2015 dropouts

At Byron Day HS they hope to fix the problem by adding new programs to make learning more fun. There are also many after school programs. The new programs mean they had to hire one new teacher for every 50 students. There are 200 students in the 2016 graduating class.

1. How many teachers did Byron Day HS hire?

Brad Knight HS decided to identify the 30% of students most at risk of dropping out. Then hire teachers to mentor them. One teacher can mentor up to 25 at-risk students. There are 160 students in the 2016 graduating class.

2. How many students did Brad Knight HS identify as most at risk of dropping out?

3. How many teachers did they hire?

The goal at both schools is to cut the percentage of dropouts to 10% or less. The programs at both schools began in 2015-2016. The dropout data for that year are shown in Table 2.

Byron Day High School			**Brad Knight High School**		
Started	**Dropped out**	**Dropout %**	**Started**	**Dropped out**	**Dropout %**
200	19		160	18	

Table 2: 2015-2016 dropouts

Both programs cut the number of dropouts. There were only 19 dropouts in Byron Day's graduating class of 2016. There were only18 dropouts in Brad Knight's graduating class of 2016.

4. Find the percentage of dropouts for each school for 2015-2016.

5. Did either school meet the goal of 10% or less dropouts?

6. Which school improved the most?

7. What evidence is there that the Byron Day program was more effective?

8. What evidence is there that the Brad Knight program was more effective?

Project idea:

Students are asked to investigate the graduation rate for your local high school for the past 5 years. Graph their findings. Write a brief paragraph with facts obtained from analyzing the data.

Practice Problems

Use Table 3 for questions #1-4.

Year	Jan	Feb	Mar	Apr	May	Jun	Jul	Aug	Sep	Oct	Nov	Dec
2006	4.7	4.8	4.7	4.7	4.6	4.6	4.7	4.7	4.5	4.4	4.5	4.4
2007	4.6	4.5	4.4	4.5	4.4	4.6	4.7	4.6	4.7	4.7	4.7	5.0
2008	5.0	4.9	5.1	5.0	5.4	5.6	5.8	6.1	6.1	6.5	6.8	7.3
2009	7.8	8.3	8.7	9.0	9.4	9.5	9.5	9.6	9.8	10.0	9.9	9.9
2010	9.8	9.8	9.9	9.9	9.6	9.4	9.4	9.5	9.5	9.4	9.8	9.3
2011	9.1	9.0	9.0	9.1	9.0	9.1	9.0	9.0	9.0	8.8	8.6	8.5
2012	8.3	8.3	8.2	8.2	8.2	8.2	8.2	8.1	7.8	7.8	7.7	7.9
2013	8.0	7.7	7.5	7.6	7.5	7.5	7.3	7.3	7.3	7.2	6.9	6.7
2014	6.6	6.7	6.7	6.2	6.2	6.1	6.2	6.2	6.0	5.7	5.8	5.6
2015	5.7	5.5	5.5	5.4	5.5	5.3	5.3	5.1	5.1	5.0	5.0	5.0
2016	4.9	4.9	5.0	5.0	4.7	4.9	4.9	4.9	5.0			

Table 3: Unemployment rates from U.S. Bureau of Labor Statistics data 2006-2016

By BLS definitions, the labor force is the following: "Included are persons 16 years of age and older residing in the 50 States and the District of Columbia who are not inmates of institutions (for example, penal and mental facilities, homes for the aged), and who are not on active duty in the Armed Forces."

1. If the labor force in March of 2010 consisted of 153,889,000 workers, how many would be considered unemployed?

2. Did the unemployment rate increase or decrease from May of 2010 to June of 2010? What percentage of the labor force gained or lost employment?

3. If the labor force remained at 153,889,000 in May 2010, how many people gained or lost jobs in June 2010? Explain your thinking.

4. Draw a line graph of the data from 2009. Identify between which months were the largest increases in the unemployment rate and which months were the largest decreases in the unemployment rate.

Use Table 4 for questions #5-7[1].

Table 1. Civilian labor force, by age, gender, race, and ethnicity, 1990, 2000, 2010, and projected 2020

[Numbers in thousands]

Group	Level				Change			Percent change			Percent distribution				Annual growth rate (percent)		
	1990	2000	2010	2020	1990–2000	2000–2010	2010–2020	1990–2000	2000–2010	2010–2020	1990	2000	2010	2020	1990–2000	2000–2010	2010–2020
Total, 16 years and older	125,840	142,583	153,889	164,360	16,743	11,306	10,471	13.3	7.9	6.8	100.0	100.0	100.0	100.0	1.3	0.8	0.7
Age, years:																	
16 to 24	22,492	22,520	20,934	18,330	28	–1,586	–2,604	.1	–7.0	–12.4	17.9	15.8	13.6	11.2	.0	–.7	–1.3
25 to 54	88,322	101,394	102,940	104,619	13,072	1,546	1,679	14.8	1.5	1.6	70.2	71.1	66.9	63.7	1.4	.2	.2
55 and older	15,026	18,669	30,014	41,411	3,643	11,345	11,397	24.2	60.8	38.0	11.9	13.1	19.5	25.2	2.2	4.9	3.3
Gender:																	
Men	69,011	76,280	81,985	87,128	7,269	5,705	5,143	10.5	7.5	6.3	54.8	53.5	53.3	53.0	1.0	.7	.6
Women	56,829	66,303	71,904	77,232	9,474	5,601	5,328	16.7	8.4	7.4	45.2	46.5	46.7	47.0	1.6	.8	.7
Age of baby boomers	26 to 44	36 to 54	46 to 64	56 to 74	...	...	...	...	...	...	...	...	...	...	...	...	...

Table 4: Civilian labor force

5. What was the total labor force in 1990?

   How did the labor force change from 1990 - 2000?

   How did you find the change?

   Where is that number in the table?

6. What does the 16,743 stand for in the table?

   What numbers in the table could you use to verify that statistic?

7. What percent change does the 16,743,000 represent?

   What numbers in the table could you use to verify that statistic?

---

[1] Source: U.S. Bureau of Labor Statistics

Activity 4: Teachers' guide

Thinking through a lesson protocol

Standards:

6.RP.A.3.C: Find a percent of a quantity as a rate per 100 (e.g., 30% of a quantity means 30/100 times the quantity); solve problems involving finding the whole, given a part and the percent.

7.RP.A.3: Use proportional relationships to solve multistep ratio and percent problems.

Mathematical Practices:

MP1: Make sense of problems and persevere in solving them.

MP2: Reason abstractly and quantitatively.

MP3: Construct viable arguments and critique the reasoning of others.

Setting up the problem - Launch	
Selecting tasks/goal setting	(5 minutes) Briefly discuss or ask students in a whole group setting what they know about graduation rates/dropout rates.
Questions	Can you think of a time that percentages are used in school other than in grading? When are they used in place of a decimal or a fraction? Why do you think they are used instead of a decimal or a fraction? Can you think of another situation where 0% is the desired outcome rather than 100%?

Monitoring student work - Explore		
Strategies and misconceptions- Anticipating	**Who - Selecting and sequencing**	**Questions and statements - Monitoring**
(10 minutes) Have students read text and answer questions #1-3 individually. Share results in whole group discussion.		What are two pieces of information that you found in Table 1? (Teacher records on board). As students share out their findings for questions 1-3 make sure that they justify their reasoning.
(30-40 minutes) Then read text and answer questions #4-8 with a partner. Have the whole group share out their answers for 4 & 5 and share out strategies for finding the percent that dropped out.		
Using Mathematics Practice #3 (Construct viable arguments and critique the reasoning of others), list on the board supporting statements for student arguments.		If students have difficulty writing the argument for their solutions, they can copy down these responses and practice saying them with their partner.

Monitoring individual student work - Explore		
Strategies and misconceptions-Anticipating	**Who - Selecting and sequencing**	**Questions and statements - Monitoring**
For off-task students or for students that seem to be self-conscious about you listening to them share.		I am just listening or looking to find out how you are working on the problem. This helps me think about what we will do later.
For students that appear to be stuck. Also for when you are having a difficult time understanding their strategies.		Can you tell me a little about your reading? How would you describe the problem in your own words? What facts do you have? What information do you have in the table? Could you try it with simpler numbers?
For students that want to ask you questions, these are ways to uncover their thinking and judge to what extent you want to respond.		Tell me what you've thought about so far. What do you know? Why are you interested in more information about that? Let me say a little about that part.

Managing the discussion – Summarize	
Parts of discussion - Connecting	**Questions and statements - Connecting**
Launching the discussion: Select the problems in questions #4-8 that students are struggling with or you wish to share out.	Will team 1 start us off by sharing one way of working on this problem? Please raise your hand when you are ready to share your solution. What did you do first when you were working on this problem? Let's start by clearing up a few things about the problem. Let's list some key parts in this problem. What was unclear in the problem?
Eliciting and uncovering student strategies	Joe would you be willing to start us off? What have you found so far? Can you repeat that? Can you explain how you got that answer? How do you know? Walk us through your steps. Where did you begin? Can you show us?
Focusing on mathematical ideas	Can you explain why this is true? Does this method always work? How is Bob's method similar to Kelly's method? What do all the solutions have in common? What would happen if I changed the numbers to _____?
Encouraging interactions	Do you agree or disagree with Kahlil's idea? What do others think? Would someone be willing to repeat what Tom just said? Would anyone be willing to add on to what Sue just said?
Concluding the discussion	Can anyone tell me some of the big ideas that we learned today? How would you explain what we learned today to a 5th grader? Some of the key points from our discussion today are . . . Tomorrow we will continue our exploration of ______ beginning with the idea from today that __________.
Post lesson notes	You may wish to assign the practice problems that you feel would benefit the students.

Solutions to text questions

1. How many teachers did Byron Day HS hire?

   1 new teacher for every 50 students is the same ratio as 2 new teachers for every 100 students or 4 new teachers for every 200 students. ***4 new teachers were hired.***

   OR

New teachers	*1*	*2*	*3*	***4***	
Students	*50*	*100*	*150*	***200***	

   4 new teachers were hired.

2. How many students did Brad Knight HS identify as most at risk of dropping out?

   30% of 160 students is the same as .3 times 160 or ***48 students were identified.***

   OR

   $$\frac{30}{100} = \frac{x}{160}$$

   $30(160) = 100x$

   $4800 = 100x$

   $48 = x$

   48 students were identified.

3. How many teachers did they hire?

Mentor Teachers	*1*	***2***	*3*
Students	*25*	***50***	*75*

   2 mentor teachers are needed for 48 students.

   OR

   $$\frac{1}{25} = \frac{x}{48}$$

   $48 = 25x$

   $1.92 = x$

   Rounding up to nearest whole person, ***2 mentor teachers are needed for 48 students.***

Byron Day High School			Brad Knight High School		
Started	**Dropped Out**	**Dropout %**	**Started**	**Dropped Out**	**Dropout %**
200	19	*9.5%*	160	18	11.3%

Table 2: 2015-2016 dropouts

4. Find the percentage of dropouts for each school for 2015-2016.

Byron Day HS 19 out of 200 is the same as 9.5 out of 100 or 9.5%

Brad Knight HS 18 is what % of 160?

$$\frac{18}{160} = \frac{x}{100}$$

1800 = 160x

11.25 = x

Byron Day HS had 9.5% dropouts.

Brad Knight HS had 11.3% dropouts.

5. Did either school meet the goal of 10% or less dropouts?

Yes, Byron Day HS had 9.5% dropouts which is less than 10%.

6. Which school improved the most?

Byron Day HS went from 21.1% dropouts to 9.5% dropouts, a decrease of 11.6%.

Brad Knight HS went from 25% dropouts to 11.3% dropouts, a decrease of 13.7%.

7. What evidence is there that the Byron Day HS program was more effective?

Looking at the number of graduates, Byron Day graduated 39 more students in 2016 than in 2015. Brad Knight only graduated 7 more students.

8. What evidence is there that the Brad Knight program was more effective?

Byron Day HS decreased the number of dropouts from 38 to 19, a 50% decrease, while Brad Knight HS decreased the number of dropouts from 45 to 18, an 82% decrease in the number of dropouts.

Solutions to Practice Problems

1. If the labor force in March of 2010 consisted of 153,889,000 workers, how many would be considered unemployed?

   Unemployment rate was 9.9%. 0.099×153,889.000 = ***15,235,011 unemployed workers.***

2. Did the unemployment rate increase or decrease from May of 2010 to June of 2010?

   Decreased by .2%

   What percentage of the labor force gained or loss employment?

   0.2% gained employment. If the unemployment rate drops, more people from the labor force are employed.

3. If the labor force remained at 153,889,000 in May 2010, how many people gained or lost jobs in June 2010? Explain your thinking.

   0.002 × 153,889,000 = 307,778 people gained jobs
   OR
   .096 × 153,889,000 = 14,773,344
   .094 × 153,889,000 = 14,465,566
   14,773,344 - 14,465,566 = ***307,778 people gained jobs***

4. Draw a line graph of the data from 2009. Identify between which months were the largest increases in the unemployment rate and which months were the largest decreases in the unemployment rate.

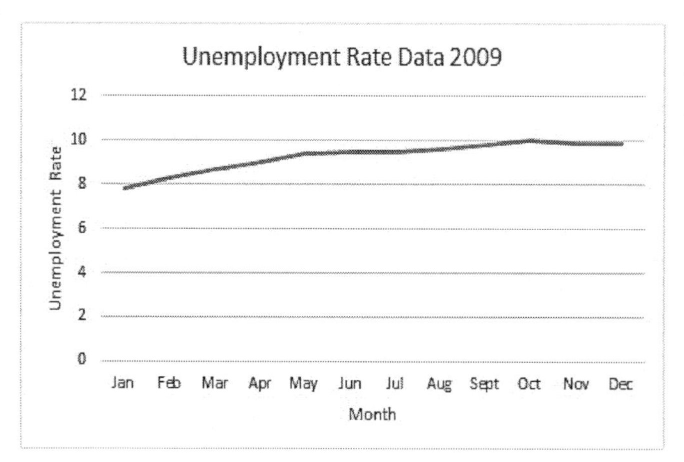

Largest increases from January to May 2009, continued to increase June to October, and decreases October to December 2009.

5. What was the total labor force in 1990? *125,840,000*
   How did the labor force change from 1990 – 2000?

   142,583,00 - 125,840,000 = 16,743,000. The labor force increased by 16,743,000 workers.

   How did you find the change?

   Subtracted labor force in 1990 from labor force in 2000.

   Where is that number in the table?

   Under column Change 1990-2000; row Total, 16 years and older.

6. What does the 16,743 stand for in the table?

   It represents the number of workers added to the labor force from 1990 to 2000 in units of thousands. The actual estimate is therefore 16,743,000.

   What numbers in the table could you use to verify that statistic?

   Subtract 125,840 from 142,583

7. What percent change does the 16,743,000 represent?

   13.3%

   What numbers in the table could you use to verify that statistic?

   16,743 / 125,840 = 0.133

Activity 5: Military Special Ops

Mathematical Goals

The student will use percentages to compare failure rates and passing rates. Students will also explore the commutative and associative properties of multiplication.

In part I the student will:

- Read a scenario and use data to complete tables
- Perform operations with percentages
- Use the commutative and associative properties of multiplication
- Calculate different percentages
- Work with percentages in a meaningful context

In part II, appropriate for students who have been introduced to basic algebra, the student will:

- Transition into the use of an algebraic expression
- Perform an economic analysis and make a decision
- Solve algebraic equations

Before the lesson (5-10 minutes)

Put the paper and pencil down and practice some mental mathematics.

Number talk possibilities:

Select two or three depending on student abilities.

- 25% of students completed the activity. What percent did not complete it?
- 45% of students did not complete the activity. What percent did complete it?
- $8 \times 15 = \_\_\_\_ \times 8$

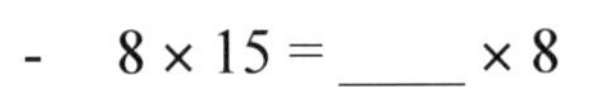

- $8 \times 3 \times 5 = 8 \times \_\_\_\_ \times 3$
- $8 \times (5 \times 3) = (8 \times \_\_\_) \times 3$
- $(6 \times \_\_\_) \times 3 = 6 \times (4 \times 3)$

Special ops – Percentage completion

Part I.

A Marine special operations force is planning a mission. They need 36 Marines specially trained for it. Not all of the Marines that start the special training will finish it. A group of 60 Marines has volunteered for this mission.

The special training takes two weeks. The first week is survival training. It is very demanding. Usually, 25% of the volunteers fail this training. The second week is cultural training when they also learn to speak the local language. Usually, 40% of volunteers fail this training.

Captain Frank McDonald and Colonel Cheryl Sanders are in charge of planning and completing the mission. They are worried that there may not be enough volunteers left after the 2 weeks of special training to have at least 36 Marines for the mission. Colonel Sanders does some figuring. She assumes 25% of the 60 volunteers will not pass survival training.

To find out how many Marines would pass and move on to cultural training, she first finds 25% of 60:

$$25\% \text{ of } 60 = 0.25 \times 60 = 15 \text{ Marines}$$

But, this is the number who will fail, and she needs to know how many will be left after week 1. There would be (60 –15) or 45 Marines remaining after week 1. The 45 Marines who are left will go on to cultural training.

1. What percent should Colonel Sanders assume will fail cultural training?

2. How many Marines is that?

3. How many of the 60 volunteers will be left after the second week?

Captain McDonald thinks changing the order of the training might produce the needed 36 Marines.

4. If the order of the training is changed, what percent should Colonel Sanders assume will fail the first week of training? What percent should she assume will fail the second week?

5. If the order of training is changed, how many Marines will be left after the first week of training? How many will be left after the second week?

Captain McDonald was surprised that the end result was exactly the same. More Marines failed the first week. Fewer failed the second week. But, the total number of failures was the same. Colonel Sanders was not surprised. She explained why the order did not matter.

For each week, they began by finding the number of Marines who do not pass. Then they subtracted this number from the total to find the number remaining. The Colonel said it would be easier to work with the percent who pass each week. If 40% fail, then 60% pass. Similarly, if 25% fail, then 75% pass.

$$0.60 \times 60 = 36$$

$$0.75 \times 36 = 27$$

In the second step if 36 is replaced by 0.60×60, then the equation is the same as

$$0.75 \times (0.60 \times 60).$$

Now, one of the properties of multiplication is that

$$0.75 \times (0.60 \times 60) = (0.75 \times 0.60) \times 60$$

6. What is the name of that property of multiplication?

Now, we can use another property of multiplication.

$$(0.75 \times 0.60) \times 60 = (0.60 \times 0.75) \times 60$$

7. What is the name of this second property of multiplication?

So, changing the order of training only changes the order of multiplying the two decimals. Captain McDonald wondered if they need to recruit as many as 100 volunteers to reach their goal. With her simplified approach, Colonel Sanders just substituted 100 in place of 60 in her equation.

$$(0.60 \times 0.75) \times 100 = 45$$

Yes, 100 volunteers are more than enough.

The Captain wondered if they could find the exact number of volunteers they need. The Colonel then showed him how to use algebra to do that. First, she noted that 27 out of 60 volunteers pass the two weeks of training.

Then she found what percent that is $27 \div 60 = 0.45$, and $0.45 \times 100\% = 45\%$

As a check, she pointed out that 45 of 100 volunteers complete training. This is also 45%. But where does 45% come from? It is found by multiplying the two passing rates:

$$0.60 \times 0.75 = 0.45 = 45\%$$

The Colonel suggested they organize the search for the right number of volunteers. First, she needed a variable. She let y represent the number of volunteers that start training. It is a variable because it can change. Next, she wrote an expression to represent the number who would complete training. This is just 45% of the volunteers.

Number who complete training $= 0.45y$

The Colonel had calculated that with 100 volunteers, there would be 45 who completed the training. If there were only 60 volunteers, then 27 would complete training. She used the expression to organize the information in Table 1.

Volunteers	**Number who complete training** **0.45 y**
100	0.45 *(100)* = *45*
95	
65	
60	0.45 *(60)* = *27*

Table 1: The number of volunteers who would complete training

8. Use the expression to determine how many would complete training if there were 95 volunteers. What if there were 65 volunteers? (Round to the nearest whole number.) Insert this information into Table 1.

9. Pick two more values for the number of volunteers and determine how many would complete training. Insert this information into Table 1.

10. Explain why in this context it is preferred to round down the estimated number who would complete training rather than rounding off the calculations in Table 1.

Part II: Solving an algebraic equation to determine number of volunteers

This process of evaluating different numbers of volunteers to find the right number is boring. The Colonel suggested that it is easier to create an algebraic equation and then solve the equation to determine the right number of volunteers that are needed. She simply set the expression equal to 36, the number needed. Next, she wrote an equation based on the 45% she assumes will finish the training.

$$0.45y = 36$$

Colonel Sanders then proceeded to solve the equation. The process involves working on the equation to transform into an equation in which the left-hand side is just $1y$. As long as one divides both sides of an equation by the same number, both sides of the equation are still equal. The Colonel divides both sides of the equation by 0.45.

$$\frac{0.45y}{0.45} = \frac{36}{0.45}$$

$$1y = 36 \div 0.45$$

$$y = 80$$

This shows them that 80 volunteers will be enough if the usual passing rates apply to this group of volunteers. Earlier Colonel Sanders showed that the order of training 80 volunteers would not affect the final total. Captain McDonald wondered if there was any reason to start with one type of training or the other. Colonel Sanders thinks it is always cheaper to start with the training that has the higher failure rate. The cost of a week of training is $2,000 for each Marine. No matter which training comes first, 80 Marines will undergo training the first week. The total cost is $160,000. If we start with the more difficult training only 60% of the Marines will be left after week 1. Sixty percent of 80 is 48 Marines. The second week will cost $96,000. The total cost to obtain 36 trained Marines is $256,000. This information is organized in Table 2.

Week	Marines	Cost	Pass %	Marines remaining
1 - Cultural	80	$160,000	60%	48
2 - Survival	48	$96,000	75%	36
Total Cost		$256,000		

Table 2: Cost of training, cultural training goes first

11. If the survival training is first, fill in the blanks in Table 3 to find the total cost of the training.

Week	Marines	Cost	Pass %	Marines remaining
1 - Survival	80		75%	
2 - Cultural			60%	
Total Cost				

Table 3: Cost of training, survival training goes first

12. Was Colonel Sanders right? Is it cheaper to start with the more difficult training?

Project idea:

Look for data on the passing rate for different types of specialized military training. What percent of soldiers make it through Navy Seal training or Army Ranger training? What percent make it through pilot training? Are there different parts of the training that result in higher failure rates?

Practice problems

Omega 4 is looking to hire graphic artists. From past experience, they know that 40% of new recruits will fail their required Basic Skills Math Test. Typically 20% will not complete their required computer programming training successfully before being hired as a graphic artist.

1. If Omega 4 currently has 50 new recruits, how many will not pass the Basic Skills Math Test?

2. How many of the 50 new recruits will pass the Basic Skills Math Test?

3. How many of the new recruits that passed the math test will not successfully complete the computer programming training?

4. How many will successfully complete the training to be hired as graphic artists?

5. Omega 4 had the new recruits take the computer programming training first. Those that passed then took the Basic Skills Math Test. How many will be available to hire as graphic artists? Show your thinking.

6. Write an expression for any given number of new recruits to show how many will be hired as graphic artists.

7. If Omega 4 wants to hire 60 graphic artists, how many new recruits do they need?

8. Marvelous Entertainment is a developer of video games. It requires applicants to pass both a math and a reading test to be eligible to be hired by the company. In the past 75% on average have passed the math test and 80% passed the reading test. What percentage will pass both tests?

9. Does it make a difference if Marvelous Entertainment administers the math test first and then the reading test or if they administer the reading test first and then the math test? Explain your reasoning.

10. Write an equation to solve for the number of applicants needed so that Marvelous Entertainment will be able to hire 25 people.

Activity 5: Teachers' guide

Thinking through a lesson protocol

Standards:

3.OA.B.5: Apply properties of operations as strategies to multiply and divide.[2]

Examples:

If $6 \times 4 = 24$ is known, then $4 \times 6 = 24$ is also known. (Commutative property of multiplication.) $3 \times 5 \times 2$ can be found by $3 \times 5 = 15$, then $15 \times 2 = 30$, or by $5 \times 2 = 10$, then $3 \times 10 = 30$. (Associative property of multiplication.)

Knowing that $8 \times 5 = 40$ and $8 \times 2 = 16$, one can find 8×7 as $8 \times (5 + 2) = (8 \times 5) + (8 \times 2) = 40 + 16 = 56$. (Distributive property.)

6.RP.A.3.C: Find a percent of a quantity as a rate per 100 (e.g., 30% of a quantity means 30/100 times the quantity); solve problems involving finding the whole, given a part and the percent.

7.RP.A.3: Use proportional relationships to solve multistep ratio and percent problems.

Mathematical Practices:

MP1: Make sense of problems and persevere in solving them.

MP2: Reason abstractly and quantitatively.

MP3: Construct viable arguments and critique the reasoning of others.

MP7: Look for and make use of structure.

Setting up the problem - Launch	
Selecting tasks/goal setting	(5 minutes) Briefly discuss or ask students in a whole group setting about what special programs they volunteered to be a part of and what special training they had to take and/or special test they had to pass for entry into the program.
Questions	What percent of the students in your school are in the band? Is there a minimum number necessary? Has there ever not been a band because of too few students?

<table>
<tr><th colspan="3">Monitoring student work – Explore</th></tr>
<tr><td colspan="3">Part I</td></tr>
<tr><td>Strategies and misconceptions- Anticipating</td><td>Who - Selecting and sequencing</td><td>Questions and statements - Monitoring</td></tr>
<tr><td>(10 Minutes) Read the text stopping at Question #1.
Have the students restate what information they learned from the reading about the situation.</td><td></td><td rowspan="5">If students have difficulty writing the argument for their solutions, they can copy down these responses and practice saying them with their partner.</td></tr>
<tr><td>(5 Minutes) Students answer Questions # 1-3 individually. Check with a partner to see that all students have the same answers.</td><td></td></tr>
<tr><td>(5 Minutes) Students continue reading text and answer Questions #4 & #5 individually. Check with a partner to see that all students have the same answers.
The teacher should ask students what they notice.</td><td></td></tr>
<tr><td>(5 Minutes) Continue to read the text and answer Questions #6 & #7.</td><td></td></tr>
<tr><td>Have students read the text stopping before Question #8.
Check for understanding - do they know where the 45% completion rate comes from and the use of the variable “y”.</td><td></td></tr>
<tr><td colspan="3">Part II</td></tr>
<tr><td>Have students read the text. This section focuses on solving the algebraic equation.
Answer Questions #11 & #12.</td><td></td><td>Depending on students’ prior knowledge of solving algebraic equations the time spent on this section may vary. Spend more time if this is their first exposure, less if they are comfortable with solving simple equations.</td></tr>
</table>

Monitoring individual student work - Explore		
Strategies and misconceptions - Anticipating	**Who - Selecting and sequencing**	**Questions and statements - Monitoring**
For off-task students or for students that seem to be self-conscious about you listening to them share.		I am just listening or looking to find out how you are working on the problem. This helps me think about what we will do later.
For students that appear to be stuck. Also for when you are having a difficult time understanding their strategies.		Can you tell me a little about your reading? How would you describe the problem in your own words? What facts do you have? Could you try it with simpler numbers?
For students that want to ask you questions, these are ways to uncover their thinking and judge to what extent you want to respond.		Tell me what you've thought about so far. What do you know? Why are you interested in more information about that? Let me say a little about that part.

Managing the discussion – Summarize	
Parts of discussion - Connecting	**Questions and statements - Connecting**
Launching the discussion: Select the problems in questions #8-12 that students are struggling with or you wish to share out.	Will team 1 start us off by sharing one way of working on this problem? Please raise your hand when you are ready to share your solution. What did you do first when you were working on this problem? Let's start by clearing up a few things about the problem. Let's list some key parts in this problem. What was unclear in the problem?
Eliciting and uncovering student strategies	Joe would you be willing to start us off? What have you found so far? Can you repeat that? Can you explain how you got that answer? How do you know? Walk us through your steps. Where did you begin? Can you show us?
Focusing on mathematical ideas	Can you explain why this is true? Does this method always work? How is Bob's method similar to Kelly's method? What do all the solutions have in common? What would happen if I changed the numbers to _____?
Encouraging interactions	Do you agree or disagree with Kahlil's idea? What do others think? Would someone be willing to repeat what Tom just said? Would anyone be willing to add on to what Sue just said?
Concluding the discussion	Can anyone tell me some of the big ideas that we learned today? How would you explain what we learned today to a 5th grader? Some of the key points from our discussion today are . . . Tomorrow we will continue our exploration of ______ beginning with the idea from today that __________.
Post lesson notes	You may wish to assign the practice problems that you feel would benefit the students. If students have had minimal exposure to reading tables, it is strongly suggested that Practice Problem #1 be teacher lead.

Solutions to text questions

1. What percent should Colonel Sanders assume will fail cultural training?

   40%

2. How many Marines is that?

   40% of the 45 remaining Marines is 0.4×45 *or 18 Marines*

3. How many of the 60 volunteers will be left after the second week?

   $45 - 18$ *is 27 Marines*

4. If the order of the training is changed, what percent should Colonel Sanders assume will fail the first week of training?

   40%

   What percent should she assume will fail the second week?

   25%

5. If the order of training is changed, how many Marines will be left after the first week of training?

   40% of the 60 volunteers or $0.4 \times 60 = 24$ *Marines that fail.* $60 - 24$ *or 36 Marines will remain.*

   How many will be left after the second week?

   25% of the 36 remaining Marines or $.25 \times 36 = 9$ *will fail. This means that* $36 - 9$ *or 27 Marines will be left.*

6. What is the name of that property of multiplication?

   Associative Property of Multiplication

7. What is the name of this second property of multiplication?

   Commutative Property of Multiplication

8. Use the algebraic expression to determine how many would complete training if there were 95 volunteers. What if there were 65 volunteers? (Round down to the nearest whole number.) Insert this information into Table 1.

9. Pick two more values for the number of volunteers and determine how many would complete training. Insert this information into Table 1.

   Answers vary. We picked 70 and 90.

Volunteers	Number who complete training $0.45\,y$
100	0.45(100) = 45
95	0.45(95) = *42.75→43*
90	0.45(*90*) = *40.5→40*
70	0.45(*70*) = *31.5 →31*
65	0.45(65) = *29.25 →29*
60	0.45 (60) = 27

Table 1: The number of volunteers who would complete training

10. Explain why in this context it is preferred to round down the estimated number who would complete training rather than rounding off the calculations in Table 1.

    The final goal was to have at least 36 complete training. Thus any number less than 36 does meet the needs of the mission. In addition, the value 45% is only an estimate of what percent would complete training. It is better to be conservative and require more volunteers. For example, 79(0.45) is 35.55. With 79 volunteers, it is not possible to have exactly 45% complete training.

11. If the order is changed, fill in the blanks in Table 2 to find the total cost of the training.

Week	Marines	Cost	Pass %	Marines remaining
1 - Survival	80	*$160,000*	75%	60
2 - Cultural	*60*	*$120,000*	60%	36
Total Cost		*$280,000*		

Table 3: Cost of training, survival training first

Cost is calculated by multiplying $2,000 times the number of Marines.

Marines Remaining is calculated by multiplying the Pass % by the number of Marines.

Total Cost is determined by adding the costs from Week 1 and 2.

12. Was Colonel Sanders right? Is it cheaper to start with the more difficult training? *Yes*

    $280,000 - $256,000 = $24,000. It is $24,000 cheaper to start with the more difficult training.

Solutions to practice problems

1. If Omega 4 currently has 50 new recruits, how many will not pass the Basic Skills Math Test?

   *40% of 50 will not pass the Basic Skills Math Test. 0.40 * 50 = 20 will not pass the Basic Skills Math Test*

2. How many of the 50 new recruits will pass the Basic Skills Math Test?

   50 - 20 = 30 will pass the Basic Skills Math Test.

3. How many of the new recruits that passed the math test will not successfully complete the computer programming training?

   20% of 30 = will not complete the computer programming training. 0.20 (30)= 6 will not complete the computer programming training.

4. How many will successfully complete the training to be hired as graphic artists?

   30 - 6 = 24 will successfully complete the training to be hired as graphic artists.

5. Omega 4 had the new recruits take the computer programming training first. Those that passed then took the Basic Skills Math Test. How many will be available to hire as graphic artists? Show your thinking.

   50 new recruits multiplied by the 80% who pass computer programming times 60% who pass Math Test = $50 \times 0.80 \times 0.60 = 24$ *will successfully pass both tests and hire in as graphic artists*

6. Write an expression for any given number of new recruits to show how many will be hired as graphic artists.

   $0.80 \times 0.60 = 0.48$ *OR 48% will complete*

   Expression is ***.48n*** *where n is the number of new recruits*

7. If Omega 4 wants to hire 60 graphic artists, how many new recruits do they need?

   $0.48n = 60$

   $n =$ ***125 new recruits are needed***

8. Marvelous Entertainment is a developer of video games. It requires applicants to pass both a math and a reading test to be eligible to be hired by the company. In the past 75% on average have passed the math test and 80% passed the reading test. What percentage will pass both tests?

*$0.75 \times 0.80 = 0.60$ OR **60%***

9. Does it make a difference if Marvelous Entertainment administers the math test first and then the reading test or if they administer the reading test first and then the math test? Explain your reasoning.

   *$0.75 \times 0.80 = 0.60$ OR **60%** and $0.80 \times 0.75 = 0.60$ OR **60%** Multiplication is commutative so it does not matter if they administer the math test or the reading test first.*

10. Write an equation to solve for the number of applicants needed so that Marvelous Entertainment will be able to hire 25 people.

    $0.60n = 25$

Activity 6:
Growing Lawn Service
Compound Percentages

Mathematical goals

The student will use percentages in determining markups and rates of growth.

The student will:

- Read a scenario and complete data presented in table format
- Perform operations with percentages
- Change fractions to decimals to percentages
- Calculate percent increase, markup, and rate of growth
- Transition into the use of algebra to answer a question
- Work with percentages in a meaningful context
- Use whole-number exponents in expressions
- Solve real-life and mathematical problems using numerical and algebraic expressions and equations

Before the lesson (5-10 minutes)

Put the paper and pencil down and practice some mental mathematics.

Number talk possibilities:

Select two or three depending on student abilities.

- What is 200% of 25?
- What is 150% of 60?
- 60 is _?_ % of 40.
- 40 is _?_ of 60.

Growing a business – Compound percentage

In all of our examples so far, the denominator was a total. The numerator was part of that total. With grading, the denominator was the total number of questions. In the NBA example, the denominator was the total number of free throws. In the dropout example, the denominator was the total number of seniors. In the training example, the denominator was the total number of volunteers. The numerator was always part of the denominator. It could not be larger than the denominator.

Sometimes a numerator can be larger than the denominator. Then percentages larger than 100% make sense. For example, businesses compare selling price to cost. A large company like Target might buy sneakers for $40 a pair. They might sell a pair of those sneakers for $100. This is $60 more than the sneakers cost. The difference between the cost of a product and the price that it sells for is called the ***markup***. In this case, the markup from a cost of $40 is $60. This markup is 150% of the cost. This time, the denominator of $40 is a basis for comparing things. The numerator, which is the markup, is not a part of the cost. It can be more or less than the cost. Therefore, the markup can be less than 100%, 100%, or more than 100% of the cost.

1. Show why a $60 markup on a cost of $40 is a 150% markup.

2. In the sneaker example, how much would a 100% markup be?

3. Give an example of a markup that is less than 100% of the sneaker cost[1].

---

[1] For students interested about costs and pricing, here is a link: http://solecollector.com/news/2014/12/how-much-it-costs-nike-to-make-a-100-shoe

The money a company receives from sales is called ***revenue***. Companies like to watch how their revenue changes. The change in a company's revenue is called its ***rate of growth***. When a company is growing, the rate of growth might be larger than 100%. In 2010, Apple Computer's revenue was $65.2 billion. In 2011, it was $108.3 billion. That was $43.1 billion more in one year[2].

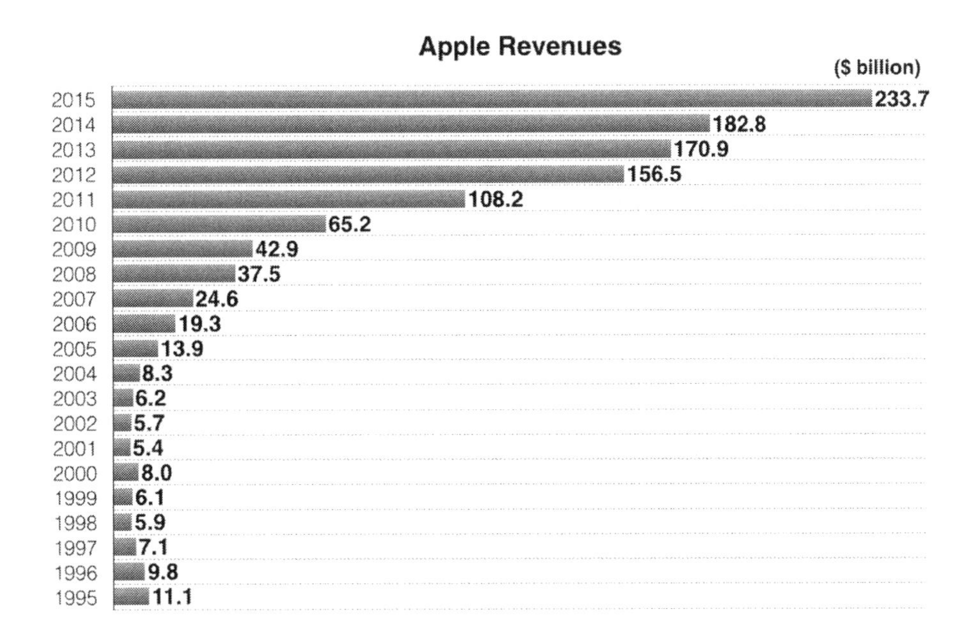

4. What was Apple's rate of growth for that year?

In 2012, Apple's revenue grew to $156.5 billion.

5. What was Apple's one-year rate of growth in 2012 compared to 2011?

6. What was Apple's two-year rate of growth in 2012 compared to 2010?

In the next two years, Apple's revenue grew from $156.5 billion in 2012 to $183.2 billion in 2014. But, some people who bought Apple stock as an investment were worried.

7. Use growth rates to explain why some investors worried about Apple's growth from 2012 to 2014.

---

[2] Adapted from Revenuesandprofits.com

Growing a lawn and garden service

Darnell and Bernice Green are co-owners of a lawn and garden service. They mow lawns and plant bushes, flowers and vegetables for area homeowners. During the first week in March, they called all of last year's customers. Fifty customers agreed to use their service again this year.

Darnell's cousin Freddie helps companies find new customers. He starts with a list of possible customers. He told Darnell that he has a long list of homeowners. He says he will contact 50 potential customers each week. His sales pitch works 30% of the time.

8. How many new customers does Freddie think he can attract in week 1? How many customers will they have at the end of week 1?

9. How many new customers does Freddie think he can attract in week 2? How many customers will they have at the end of week 2?

Bernice's friend Grace helps lawn services get new customers. She uses "word-of-mouth" from satisfied customers. She says she has helped others increase their number of customers by about 20% each week.

10. How many new customers does Grace think she can bring in the first week? How many customers will they have at the end of week 1?

11. If Grace was right about the first week, how many customers does she think she can bring in the second week?

Darnell and Bernice are looking at both proposals. They think there might only be four more weeks, but no more than six more weeks to sign up new customers. After that, all neighborhood homeowners will have signed lawn and garden service contracts.

	Freddie - using a list			**Grace -using word-of-mouth**		
End of week	**Start of week**	**Add 30% of 50**	**End of week**	**Start of week**	**Add 20% of previous total**	**End of week**
1	50	15	65	50	10	60
2	65	15	80	60	12	72
3						
4						
5						
6						

Table 1: Comparison of the number of customers with different marketing

Bernice is several years older than Darnell. She has already completed two years of algebra. She did well in both classes. Darnell was still in his first year of algebra. He was not yet confident about using algebra to solve meaningful problems. She told her brother that they could use algebra to help compare the two plans. She began with Freddie's plan. Each week he would recruit 15 new customers. That would mean at the end of week 2, the number of customers would be 80. This is found by multiplying 15 new customers per week by 2 weeks and adding it to the original 50 customers.

$$50 + 15 \times 2 = 80$$

At the end of week 3, the total would be 95.

$$50 + 15 \times 3 = 95$$

The total at the end of any week can be determined with an algebraic equation.

Let x = number of weeks

Number of customers with Freddie's plan= $50 + 15x$

To be sure Darnell understood her idea, she asked him to use the algebraic expression to complete the section in Table 1 under the heading "Freddie Using a List".

12. Complete this section of Table 1.

13. If they use Freddie, how many customers will they have after 5 weeks?

14. If they use Freddie, how many weeks will it take for them to at least double the number of customers?

Bernice told Freddie that she could also write an algebraic expression for Grace's "Word-of-Mouth" strategy. However, the algebraic expression is more complicated. To begin, Bernice reviews the facts. At the end of week 1, the number of customers would be 20% more than the 50 they started with.

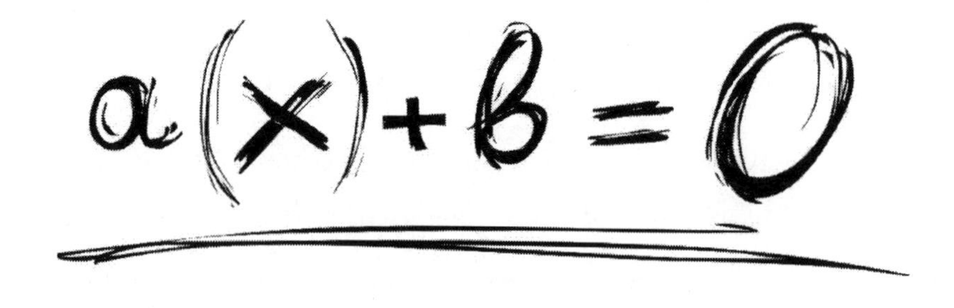

The number of customers at the end of week 1 is $50 + 0.2(50)$ or $50 + 10 = 60$

The expression $50 + 0.2 * 50$ can be rewritten using the distributive property as

$50\ (1+0.2)$ or $50(1.2) = 60$

At the end of week 2, the number of customers would be 20% more than the 60 customers at the end of week 1.

Number of customers at the end of week 2 is $60 + 0.2(60)$ or $1.2(60) = 72$

To see the pattern, we need to replace 60 with its equivalent expression 1.2(50)

$1.2(60) = 1.2(1.2 \times 50)$

You can regroup using the associative property.

$(1.2 \times 1.2)\ 50 = 1.2^2\ (50)$

Notice the value 1.2 is raised to the second power. This shows that number of customers after two weeks is found by multiplying the original 50 by 1.2 and again by 1.2.

15. Use a calculator to demonstrate that using this expression results in 72 customers at the end of week 2 with Grace's plan.

Bernice then showed how to create an algebraic expression to calculate the number of customers at the end of any week.

Let x = number of weeks

This variable also represents the number of times 50 is multiplied by the same value 1.2 The algebraic expression is

Number of customers at the end of week x is represented by 1.2^x (50)

In this case the variable x appears as the exponent. Bernice demonstrated the use of this expression for week 3.

To see the pattern, we continue into week 3. At the end of week 3, the number of customers was 20% more than the 72 customers at the end of week 2.

Number of customers at the end of week 3 is $1.2^3(50) = 1.728(50) = 86.4$

Thus, the predicted number of customers at the end of week 3 would be 86.

16. Demonstrate that you would obtain the same value starting with 72 customers at the end of week 2 and growing by 20% in week 3.

Darnell told Bernice that algebra looked interesting. However, it seemed to take as much effort as calculating the total number of new customers week by week. Bernice responded that algebra becomes more useful as the numbers grow larger. Let's assume we are most interested in which marketing plan would produce the largest number of customers at the end of six weeks.

17. Use the algebraic expression for Freddie's marketing plan to find the total number of customers after six weeks.

    Number of customers with Freddie's plan is $50 + 15x$

18. Use the algebraic expression for Grace's marketing plan to find the total number of customers after six weeks.

    Number of customers with Grace's plan is $1.2^x(50)$

Bernice thought that a graph might help Darnell see how the two plans compare over time. Figure 1 compares the two marketing plan. The x-axis is the number of weeks. The y-axis is the number customers at the end of each.

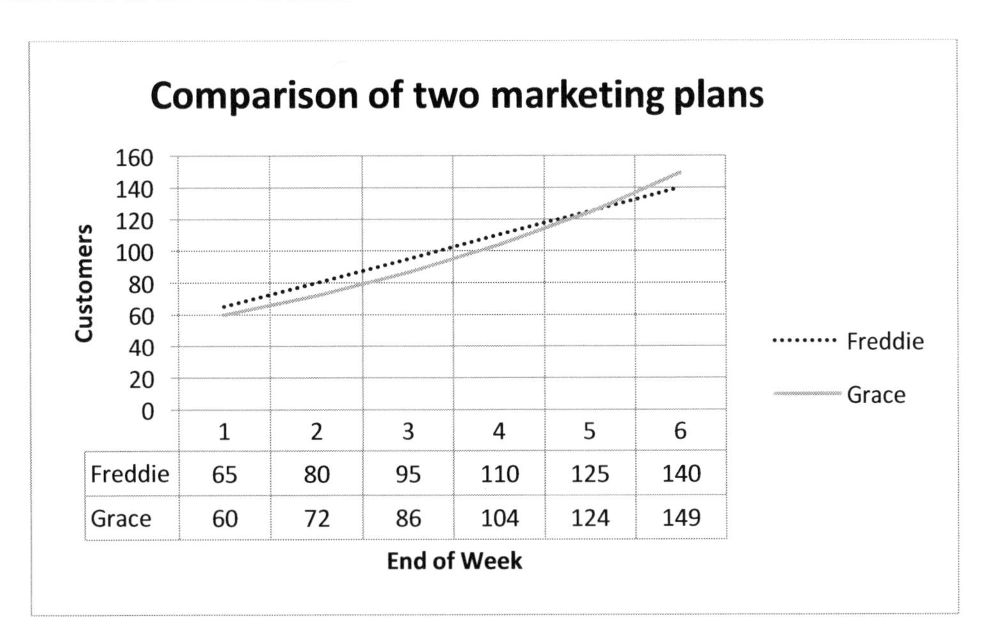

	1	2	3	4	5	6
Freddie	65	80	95	110	125	140
Grace	60	72	86	104	124	149

Figure 1: Comparison of customer growth with two marketing plans

19. Looking at the graph, when do the two proposals recruit the same number of customers?

20. If they were able to attract new customers for only four weeks, which plan is better?

21. If they were able to attract new customers for a full six weeks, which plan is better?

22. Which plan would you recommend and why?

Project idea:
Look for and interpret a graph that shows rapid growth of some app, game, or other product.

Practice problems

Chad lives in Fort Wayne, IN. He has been reading a great deal about hoverboards. Chad decides that he would like to purchase a hoverboard and has already saved $100 towards the purchase. Chad searched the web and found prices varied a great deal. The hoverboard price was affected by its maximum speed and the quality of its battery. The Swoosh hoverboard he considered cost $233.95. He was also excited by the extra capability of a Runfast hoverboard that was priced $389.50. The really fancy ones with lots of add-ons cost over $600. Chad felt that was way beyond his budget. Both his Mom and Dad were willing to pay him to work on their projects during the summer. Chad had lots of plans for the summer and could only take one of the two jobs.

	Mom's offer increase $1 per week		Dad's offer increase 10% per week	
Week	**Weekly pay**	**Total earned**	**Weekly pay**	**Total earned**
1	$20	$20	$14.00	$14.00
2	$21	$41	15.40	$29.40
3	$22	$63	$16.94	$46.34
4				
5				
6				
7				
8				
9				
10				
11				
12				
13				
14				
15				
16				

Table 2: Salary and total earnings for two different job offers

His mom was willing to pay him to grow vegetables in a large area in the sunny part of their backyard. She was planning on starting his pay at $20 for the first week. However, she wanted to encourage Chad to stay with the job throughout the summer. She offered to increase his pay by $1 each week for 16 weeks of work.

1. How much would he earn in week 4 under Mom's offer? What would his total earnings be through week 4 under Mom's offer?
2. Write an expression to determine Chad's earnings in week n. Use the expression to calculate how much Chad would earn in week 5? How much would he earn in week 16?
3. Complete Table 2 for Mom's offer.

   Chad wanted to buy the hoverboard as soon as possible. The summer months were ideal for learning all of the things the hoverboard could do.
4. When would Chad have enough money to purchase the Swoosh Hoverboard?
5. Would he earn enough money over the summer to purchase the Runfast?

Chad was showing his friend Celine that he soon would be able to share his new hoverboard with her. Celine looked over the numbers. She pointed out that Chad had forgotten that Indiana has a 7% sales tax.

6. When would Chad have enough money to purchase the Swoosh hoverboard now that he realizes there is a sales tax?

His Dad was interested in Chad helping with numerous chores around the house and yard. These include mowing the lawn, trimming the bushes, fixing the gutters, cleaning out the garage and basement. His Dad was really into percentages. He offered to start Chad off with just $14 the first week. However, each week after week 1, Chad's pay would increase by 10%. His Dad wrote out the algebraic expression for Chad's pay each week.

Weekly pay equals $(14)1.1^{(x-1)}$

7. Why is the exponent *(x-1)* and not *x*?

8. What would Chad be paid in week 1? In week 5? In Week 16?

9. Complete Table 2 for Dad's offer. When would Chad have enough money to purchase the Swoosh Hoverboard? (Do not forget sales tax.)

10. Would he earn enough money over the summer to purchase the Runfast?

11. Graph the weekly pay for each of the two salary offers.

12. Where do the two graphs intersect? What does this represent?

13. During which weeks would his Mom's offer be higher? During which weeks would his Dad's offer be higher?

14. Graph the total amount earned pay for each of the two salary offers.

15. Where do the two graphs intersect? What does this represent?

16. When do Chad's total earnings with his Dad's offer first exceed his total with his Mom's offer?

17. Which job offer would you recommend Chad take and why?

Activity 6: Teachers' guide

Thinking through a lesson protocol

Standards:

5.NF.B.3: Interpret a fraction as division of the numerator by the denominator ($a/b = a \div b$). Solve word problems involving division of whole numbers leading to answers in the form of fractions or mixed numbers, e.g., by using visual fraction models or equations to represent the problem.

6.RP.A.3.C: Find a percent of a quantity as a rate per 100 (e.g., 30% of a quantity means 30/100 times the quantity); solve problems involving finding the whole, given a part and the percent.

6.EE.A.1: Write and evaluate numerical expressions involving whole-number exponents.

7.RP.A.3: Use proportional relationships to solve multistep ratio and percent problems.

Mathematical Practices:

MP1: Make sense of problems and persevere in solving them.

MP2: Reason abstractly and quantitatively.

MP3: Construct viable arguments and critique the reasoning of others.

MP4: Model with mathematics.

MP8: Look for and express regularity in repeated reasoning.

Setting up the problem - Launch	
Selecting tasks/goal setting	(5 minutes) Briefly discuss or ask students in a whole group setting if they could give an example of retail vs wholesale pricing? Have they ever heard of companies marking-up their goods?
Questions	Have you ever heard of percentages over 100%[3]?

[3] Possible website to visit is http://solecollector.com/news/2014/12/how-much-it-costs-nike-to-make-a-100-shoe

Monitoring student work – Explore		
Strategies and misconceptions- Anticipating	**Who - Selecting and sequencing**	**Questions and statements - Monitoring**
(10 Minutes) Read the first two paragraphs and answer the three questions and discuss results.		Be sure students understand the scenario.
(20 Minutes) Continue to read paragraph on revenue and rate of growth. With a partner, answer questions #4-7. Share out whole class.		
(15 minutes) Read and discuss Growing a Lawn and Garden Service paragraphs and have students work on solving questions #8-11. Share results.		
(20 minutes) Teacher leads whole class through completing the Freddie - Using a List part of Table 1 referring to the text having students highlight expressions and equations. Students complete questions #13-14.		
(30 minutes) Teacher leads whole class through completing the Grace - Using Word-of-Mouth part of Table 1 referring to the text having students highlight expressions and equations. Students complete questions #15-18.		
(30 minutes) Look at the graph and make sense of it as a whole class. Discuss scales, axes, which graph represents Freddie and which represents Grace, etc. Complete questions #19-22.		Exponents are introduced here. Depending on students' prior experiences with exponents this may take longer and need more explicit instruction. Make sure that students understand the positions of the end of week 1, week 2, etc as being in between the vertical lines.

Monitoring individual student work - Explore		
Strategies and misconceptions - Anticipating	**Who - Selecting and sequencing**	**Questions and statements - Monitoring**
For off-task students or for students that seem to be self-conscious about you listening to them share.		I am just listening or looking to find out how you are working on the problem. This helps me think about what we will do later.
For students that appear to be stuck. Also for when you are having a difficult time understanding their strategies.		Can you tell me a little about your reading? How would you describe the problem in your own words? What facts do you have? Could you try it with simpler numbers?
For students that want to ask you questions, these are ways to uncover their thinking and judge to what extent you want to respond.		Why are you interested in more information about that? Let me say a little about that part. Tell me what you've thought about so far. What do you know?

Managing the discussion – Summarize	
Parts of discussion - Connecting	**Questions and statements - Connecting**
Launching the discussion: Select the problems in questions #8-12 that students are struggling with or you wish to share out.	Will team 1 start us off by sharing one way of working on this problem? Please raise your hand when you are ready to share your solution. What did you do first when you were working on this problem? Let's start by clearing up a few things about the problem. Let's list some key parts in this problem. What was unclear in the problem?
Eliciting and uncovering student strategies	Joe would you be willing to start us off? What have you found so far? Can you repeat that? Can you explain how you got that answer? How do you know? Walk us through your steps. Where did you begin? Can you show us?
Focusing on mathematical ideas	Can you explain why this is true? Does this method always work? How is Bob's method similar to Kelly's method? What do all the solutions have in common? What would happen if I changed the numbers to _____?
Encouraging interactions	Do you agree or disagree with Kahlil's idea? What do others think? Would someone be willing to repeat what Tom just said? Would anyone be willing to add on to what Sue just said?
Concluding the discussion	Can anyone tell me some of the big ideas that we learned today? How would you explain what we learned today to a 5th grader? Some of the key points from our discussion today are . . . Tomorrow we will continue our exploration of ______ beginning with the idea from today that ___________.
Post lesson notes	You may wish to assign the practice problems that you feel would benefit the students. If students have had minimal exposure to reading tables, it is strongly suggested that Practice Problem #1 be teacher lead.

Solutions to text questions

1. Show why a $60 markup on a cost of $40 is a 150% markup.

   Ask the question "$60 is what % of $40?" Translate to an equation 60 = x(40).

   $\frac{60}{40} = x$

   $1.5 = x$

   $1.5\ (100\%) = 150\%$

2. In the sneaker example, how much would a 100% markup be?

   Ask the question "What is 100% of $40?" Translate to an equation x = 1 (40).

   x = $40.

3. Give an example of a markup that is less than 100% of the sneaker cost.

   Since the sneakers cost $40, anything less than $40 would make the markup less than 100%. For example, if the markup was $20 (sneakers selling for $60), the markup would be $\frac{20}{40}$ *or 50%.*

4. What was Apple's rate of growth for that year?

   The amount of the growth divided by the starting amount will give us the rate of growth for a year. $\frac{43.1}{64.2} = 0.6713$

   $0.6713 \times 100\%$ *is* ***67.1% growth***

5. What was Apple's one-year rate of growth in 2012 compared to 2011?

   $156.5 - 108.3 = 48.2$

   $\frac{48.2}{108.3} = 0.4450$ *OR* ***44.5%***

6. What was Apple's two-year rate of growth in 2012 compared to 2010?

   $156.5 - 65.2 = 91.3$

   $\frac{91.3}{65.2} = 1.400$ *OR* ***140%***

7. Use growth rates to explain why some investors worried about Apple's growth from 2012 to 2014.

   The growth rate was 140% from 2010 to 2012

   $183.2 - 156.5 = 26.7$

$\frac{26.7}{156.5} = 0.1706$

The growth rate was 17.1% from 2012 to 2014

The growth rate is a much smaller percentage during the 2012-2014 years.

8. How many new customers does Freddie think he can attract in week 1? How many customers will they have at the end of week 1?

   *He will attract 30% of 50 which is **15 new customers a week**. There will be 50+15= **65 customers** at the end of week 1.*

9. How many new customers does Freddie think he can attract in week 2? How many customers will they have at the end of week 2?

   *He will again attract 30% of 50 which is **15 new customers a week**. There will be 65+15 = **80 total customers** at the end of week 1.*

10. How many new customers does Grace think she can bring in the first week? How many customers will they have at the end of week 1?

    *She will attract 20% of 50 which is **10 new customers in the first week**. There will be 50+10= **60 customers** at the end of week 1.*

11. If Grace was right about the first week, how many customers does she think she can bring in the second week?

    *At the beginning of week 2, there would be **60 customers**. She will attract 20% of 60 which is 12 new customers in the second week. There will be 60+12= **72 customers** at the end of week 2.*

12. Complete this section of Table 1.

    See Table 1 completed.

	Freddie - using a list			Grace -using word-of-mouth		
End of week	**Start of week**	**Add 30% of 50**	**End of week**	**Start of week**	**Add 20% of 50**	**End of week**
1	50	15	65	50	10	60
2	65	15	80	60	12	72
3	*80*	*15*	*95*	*72*	*14*	*86*
4	*95*	*15*	*110*	*86*	*17*	*103*
5	*110*	*15*	*125*	*103*	*21*	*124*
6	*125*	*15*	*140*	*124*	*25*	*149*

Table 1: Comparison of the number of customers with different marketing

13. If they use Freddie, how many customers will they have after 5 weeks?

    50 + 15(5) = 125

14. If they use Freddie, how many weeks will it take for them to double the number of customers?

    *After **4 weeks**, they would have **110 customers**. This more than double the number they started with.*

15. Use a calculator to demonstrate that using this expression results in 72 customers at the end of week 2 with Grace's plan.
16. Demonstrate that you would obtain the same value starting with 72 customers at the end of week 2 and growing by 20% in week 3.

    20%(72) = 14 new customers
    72+14 = 86 total customers

17. Use the algebraic expression for Freddie's marketing plan to find the total number of customers after six weeks.
    Number of customers with Freddie's plan= 50 + 15 * x

    50 + 15 (6) = 15 + 90 = 140

18. Use the algebraic expression for Grace's marketing plan to find the total number of customers after six weeks.
    Number of customers with Grace's plan = $1.2^x(50)$

    $1.2^6(50) = 2.986(50) = 149$

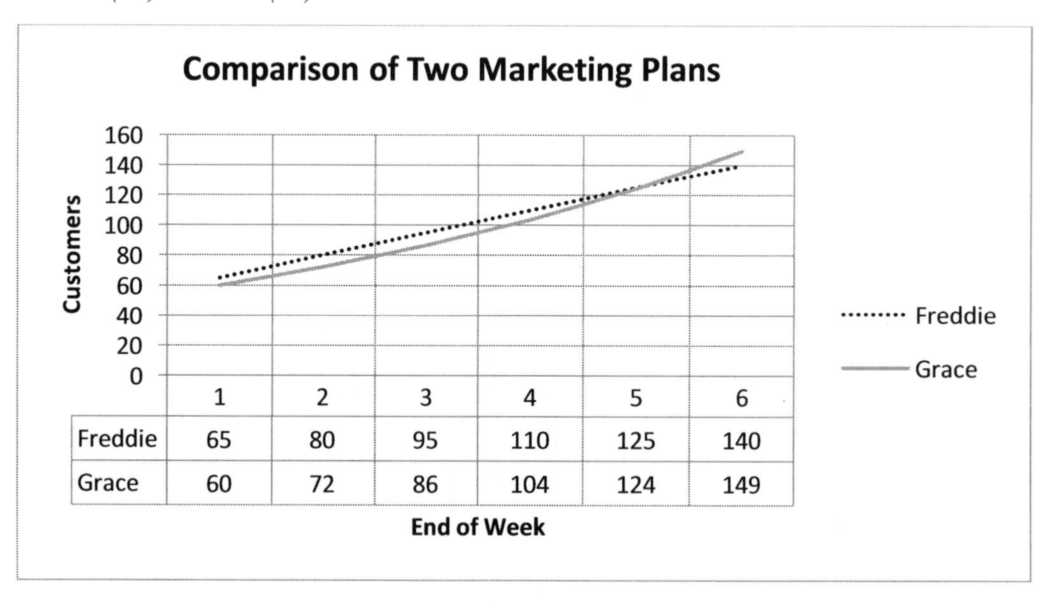

	1	2	3	4	5	6
Freddie	65	80	95	110	125	140
Grace	60	72	86	104	124	149

Figure 1: Comparison of customer growth with two marketing plans

19. Looking at the graph, when do the two proposals recruit the same number of customers?

    Approximately at the end of week 5, either plan results in a total of about 125 customers.

20. If they were able to attract new customers for only four weeks, which plan is better?

    In the first four weeks, Freddie's plan is better. Freddie's line graph is still higher on the y-axis after the first four weeks.

21. If they were able to attract new customers for a full six weeks, which plan is better?

    If they can continue marketing through the sixth week, Grace's plan is better. Grace's line graph is higher on the y-axis after week 6 weeks.

22. Which plan would you recommend and why?

    Answers will vary. If they are cautious they might assume the marketing plan will be effective for only five weeks. In that case, they would use Freddie. If they liked to take risks, they might use Grace's approach and hope marketing can continue for at least six weeks.

Solutions to practice problems

	Mom's offer increase $1 per week		Dad's offer increase 10% per week	
Week	**Weekly pay**	**Total earned**	**Weekly pay**	**Total earned**
1	$20	$20	$14.00	$14.00
2	$21	$41	$15.40	$29.40
3	$22	$63	$16.94	$46.34
4	*$23*	*$86*	*$18.63*	*$64.97*
5	*$24*	*$110*	*$20.49*	*$85.46*
6	*$25*	*$135*	*$22.54*	*$108.00*
7	*$26*	*$161*	*$24.79*	*$132.79*
8	*$27*	*$188*	*$27.27*	*$160.06*
9	*$28*	*$216*	*$30.00*	*$190.06*
10	*$29*	*$245*	*$33.00*	*$223.06*
11	*$30*	*$275*	*$36.30*	*$259.36*
12	*$31*	*$306*	*$39.93*	*$299.29*
13	*$32*	*$338*	*$43.92*	*$343.21*
14	*$33*	*$371*	*$48.31*	*$391.52*
15	*$34*	*$405*	*$53.14*	*$444.66*
16	*$35*	*$440*	*$58.45*	*$503.11*

Table 2: Salary and total earnings for two different job offers

1. How much would he earn in week 4 under Mom's offer? What would his total earnings be through week 4 under Mom's offer?

   $20 + $1 + $1 + $1 = $23 Total earnings would be $20 + $21 + $22 + $23 = $86.

2. Write an expression to determine Chad's earnings in week ***n***. Use the expression to calculate how much would he earn in week 5? How much would he earn in week 16?

   $20 + (n - 1) = Earnings in week n

   $20 + (5 - 1) = $24 earnings in week 5

   $20 + (16 - 1) = $35 earnings in week 16

3. Complete Table 2 for Mom's offer.

   See table above.

Chad wanted to buy the hover board as soon as possible. The summer months were ideal for learning all of the things the hover board could do.

4. When would Chad have enough money to purchase the Swoosh Hoverboard?

   According to the table and given that Chad has already saved $100, he will have enough money to purchase the Swoosh Hoverboard in week 6.

5. Would he earn enough money over the summer to purchase the Runfast?

   According to the table and given that Chad has already saved $100, he will have enough money to purchase the Runfast Hoverboard in week 12.

6. When would Chad have enough money to purchase the Swoosh Hoverboard now that he realizes there is a sales tax?

   $233.95(1.07) = $250.33. According to the table and given that Chad has already saved $100, he will have enough money to purchase the Swoosh Hoverboard in week 7.

Weekly pay equals $14(1.1^{(x-1)})$

7. Why is the exponent $(x-1)$ and not x?

   The first 10% increase does not occur in week one. It actually occurs in week two. That is one less than the number of weeks.

8. What would Chad be paid in week 1? In week 5? In Week 16?

   In week 1 he would be paid $14.
   In week 5 he would be paid $\$14(1.1)^{(5-1)} = \$20.49$.
   In week 16 he would be paid $\$14(1.1)^{(16-1)} = \$58.48$.

   Note: Values in the table may differ due to rounding. The values in the table were calculated by taking the previous week's earnings and adding 10%. Answers were rounded to nearest penny at each step. Rounding discrepancies occur.

9. Complete Table 2 for Dad's offer. When would Chad have enough money to purchase the Swoosh Hoverboard? (Do not forget sales tax.)

   According to the table and including the sales tax, also given that Chad has already saved $100, he will have enough money to purchase the Swoosh Hoverboard in week 8.

10. Would he earn enough money over the summer to purchase the Runfast?

    YES. If he accepts Mom's offer, he could afford the Runfast in week 13. He could also afford the Runfast in week 13 accepting Dad's offer.

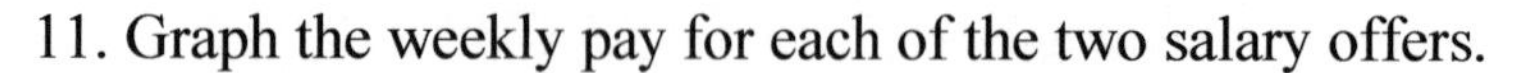

11. Graph the weekly pay for each of the two salary offers.

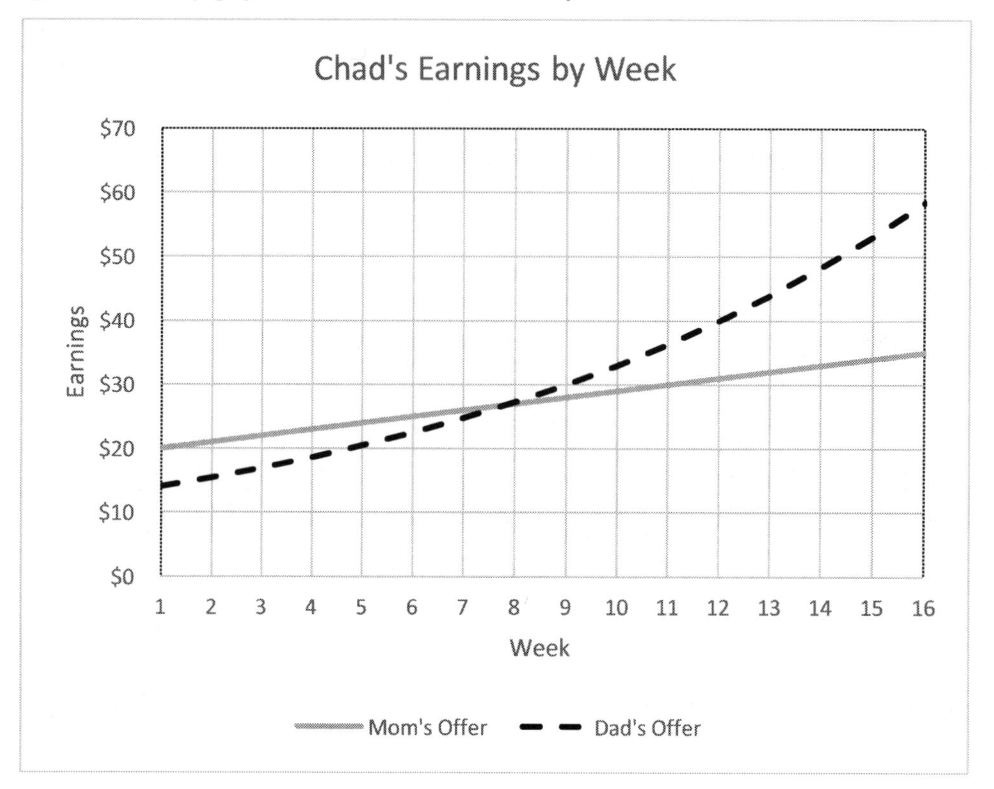

12. Where do the two graphs intersect? What does this represent?

    The two graphs appear to intersect around Week 8. This means that Mom's offer and Dad's offer will both pay Chad about the same amount for the week in Week 8.

13. During which weeks would his Mom's offer be higher? During which weeks would his Dad's offer be higher?

    Looking at the graphs, Mom's offer pays more per week during weeks 1-7; Dad's offer pays more per week during weeks 9-16.

14. Graph the total amount earned pay for each of the two salary offers.

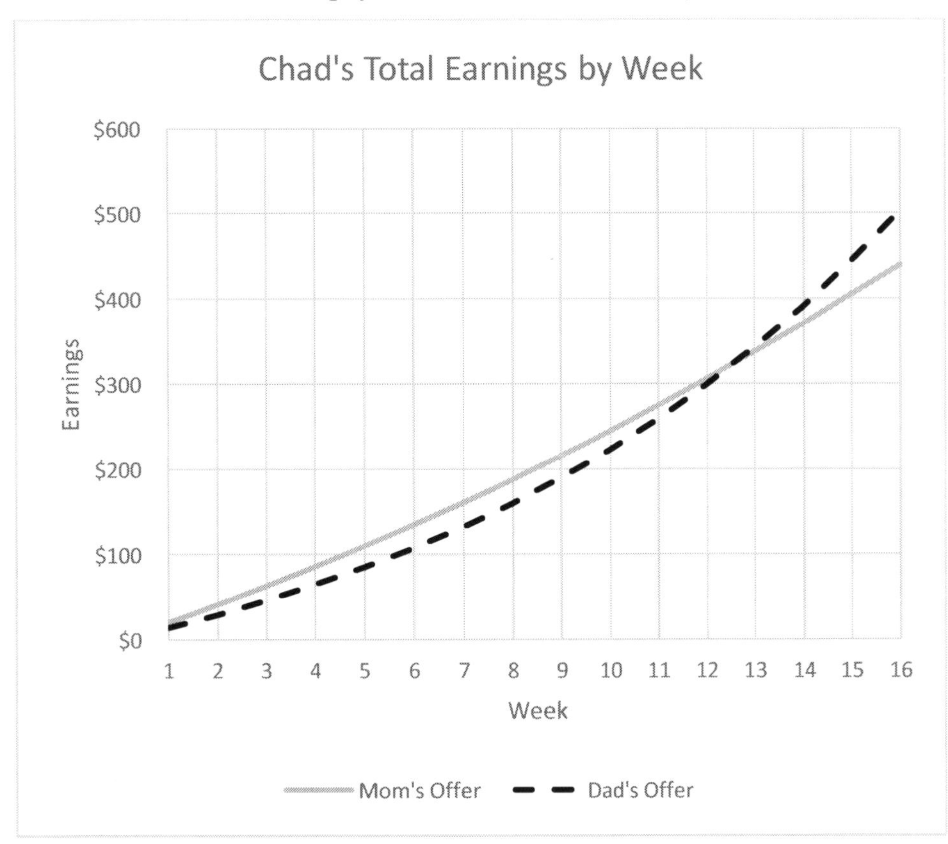

15. Where do the two graphs intersect? What does this represent?

The two graphs appear to intersect around Week 13. This means that around Week 13 the total amount of money that Chad has earning is about the same from both salary offers.

16. When do Chad's total earnings with his Dad's offer first exceed his total with his Mom's offer?

Week 13

17. Which job offer would you recommend Chad take and why?

Answers will vary. Example: If Chad chooses the Swoosh Hoverboard, he will have earned enough after Week 8 under both Mom's offer and Dad's offer. If Chad chooses the Runfast, he will have earned enough after Week 13 under both salary offers. Chad's decision could be based on the type of work he would prefer to do or the number of hours per week he feels each type of job would take him.

Activity 7:
McFadden Restaurant
Changing Revenue

Mathematical goals

The student will use percentages to explore the effects of a given percentage increase followed by the same percentage decrease in the context of owning a business.

The student will:

- Read a scenario and use data presented in tables and graphs
- Calculate different percentages
- Calculate averages of two numbers
- Transition into the use of Algebra to answer a question
- Work with percentages in a meaningful context familiar to students

Before the lesson (5-10 minutes)

Put the paper and pencil down and practice some mental mathematics.

Number talk possibilities:

Select two or three depending on student abilities.

- Increase 100 by 50%.
- Decrease 150 by 50%.
- Increase 40 by 25%.
- Decrease 50 by 25%.

McFadden Restaurant – equal percentage up and down

Ellen McDaniel owns a fast food restaurant. It is part of the McFadden Chicken Breast chain. Their specialty is a grilled chicken breast sandwich smothered in their special barbecue sauce.

Ellen's restaurant has been open for one year. She keeps track of the monthly sales. Each month, she compares her sales to the month before. She opened in January 2016. Her first month's sales were $100,000. However, her sales decreased $10,000 in February. This was a 10% decrease from the month before. The restaurant's sales increased $9,000 in March. This was a 10% increase from February. Now she was confused. In one month sales decreased by 10%. The next month sales increased by 10%. She thought the restaurant should be back to where it started. She thought sales should again be $100,000, but they weren't. They were only $99,000 in March. She wondered where the missing $1,000 was!

Month	Revenue
January	$100,000
February	$90,000
March	$99,000
April	$105,300
May	$120,750
June	$125,000
July	$135,000
August	$124,200
September	$125,000
October	$120,000
November	$118,000
December	$105,000
Total	$1,367,250

Table 1: Monthly revenue

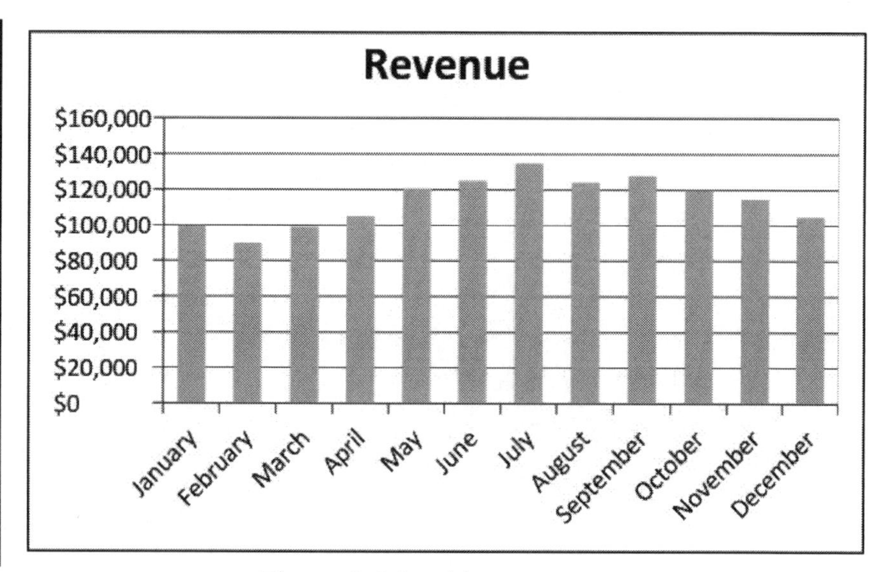

Figure 1: Monthly revenue

Understanding changes in sales

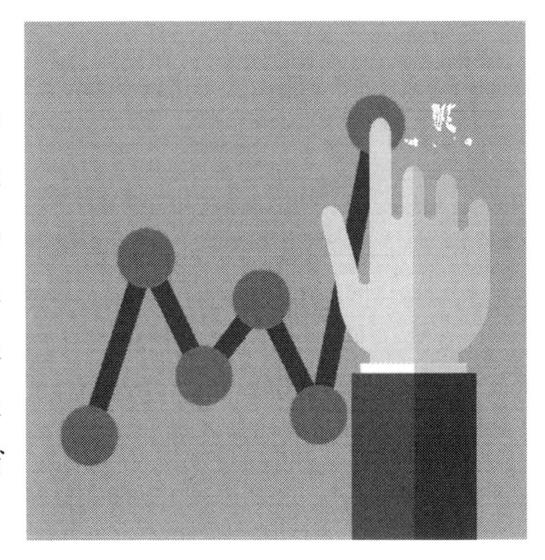

Ellen's son Daniel was learning about percentages. He learned of a common misunderstanding about percentage change. He explained that in February, the 10% decrease was 10% of the January sales. Ten percent of $100,000 is $10,000. The smaller sales in February were $90,000. Now the 10% increase in March was 10% of the February sales. Ten percent of $90,000 is $9,000.

1. How much did the restaurant sales increase from March to April?

2. What was the percent change in restaurant's sales in from March to April?

Ellen asked Daniel if the same thing would happen if the order was reversed. What if there was a percentage increase one month, and there was the same percentage decrease the next month? Daniel suggested looking at the months June through August. Sales increased from June to July by $10,000. Dividing this increase by $125,000 in sales for June equals 0.08. Then (0.08)(100%) is 8%. There was an 8% increase in sales.

From July to August, sales decreased to $124,200. This is a $10,800 decrease. To find the percentage decrease, he divided $10,800 by $135,000, the sales in July. The answer was 0.08. Then he multiplied 0.08 by 100%. This is also 8%. Thus, the percent decrease was the same as the previous month's increase. Once again, the final number $124,200 is less than the starting number of $125,000. Daniel's mother still looked confused.

Daniel decided to show his mother the mathematics behind this pattern. They took a different look at the numbers. Sales in February were $90,000. This is equal to multiplying $100,000 by 0.90. Now 0.90*100% is 90%. So, February sales were 90% of January's sales, which is 10% less. Sales increased by 10% in March. The March total is 100% of the February total + 10% more. So, March sales were 110% of February sales, and 110% = 1.10. Multiplying 1.10 by 0.9 equals 0.99. This is 0.01 or 1% less than the original value.

February Sales = 0.90 × January Sales

\$90,000 = 0.90 × \$100,000

March Sales = 1.10 × February Sales

99000= 1.10 × 90000

March sales = 1.10 × (0.90 × January Sales)

= (1.10 × 0.90) × January Sales

= 0.99 × January sales.

The associative law of multiplication allows us to regroup the multiplication. So, in the end, we multiply 0.99 by the January sales. Let's consider the summer data.

July Sales = 1.08 × June Sales

August Sales = 0.92 × July Sales

August sales = 0.92 × (1.08 × June Sales) = (0.92 × 1.08) × June Sales

August sales = 0.9936 × June Sales.

It turns out that no matter what the percentages are, increasing and decreasing by the same percentage in either order always ends up with a smaller number. Table 2 presents a few different percentage changes.

3. Fill in the correct values for the last column in the Table 2.

4. As the percent change increases, what happens to the value of I times D?

5. With a 50% increase and a 50% decrease, what percent of the original number is the final number?

Percent change	I = Increase	D = Decrease	Final value multiply I×D
5%	1.05	0.95	0.998
10%	1.10	0.90	
25%	1.25	0.75	
33%	1.33	0.67	
50%	1.50	0.50	

Table 2: The result of equal percentage increases and decreases

Now Ellen felt she understood the math. Next, Ellen thought about a more important question. Why did sales decrease by 10% in February? She thought maybe bad weather made the number of customers lower. However, February did not seem any worse than January. Daniel reminded his mother of one important fact. These data are *monthly* sales. There are three less days in February than in January. January has 31 days. Three divided by 31 is 0.097, and 0.097 × 100% = 9.7%. There are 9.7% less days in February than in January. Now Ellen was not so surprised that monthly sales went down from January to February.

Daniel explained that it's better to compare average sales per day. To find the average sales per day, Daniel explained that the monthly sales total is divided by the number of days in the month.

6. In January, what was the average sales per day?

7. What was the average sales per day in February?

8. Do you think Ellen and Daniel should worry about the difference in the average daily sales in February compared to January?

Now Ellen was not surprised by the increase in sales from February to March. March also has 31 days which is 3 more days than February.

9. Find the average sales per day in March.

10. Should Ellen and Daniel think that the restaurant's sales in January, February, and March were up-and-down, or pretty much the same?

Sales tax and contract fees

Daniel's understanding of percentages impressed Ellen. She asked him to help her with two more percentages. As part of her contract with the McFadden Chicken Breast chain, every year she must pay the company fee of 10% of all sales. In addition, she must collect 6% sales tax on all sales. This she sends to the State Sales Tax Office. Her total revenue for 2014 was $1,367,250. This included the sales taxes she collected. Ellen was going to write a check to McFadden Chicken Breast Co. for 10% of $1,367,250.

11. What was the amount of the check Ellen was going to write?

Something was bothering Ellen. She knew that the McFadden fee was supposed to be based on sales, and she knew what her total revenue was. But she also knew that her total revenue included the sales tax that she had collected. So, her total sales must have been less than her total revenue, but she didn't know how to find what her total sales were. She asked Daniel if he knew how to find the total sales.

Daniel showed Ellen how to use an equation to find the total sales. He began by saying,

Let: S = the total sales

Then, the total of the sales tax collected was 6% of $S = (0.06)(S)$.

"But what about that $1,367,250?" Ellen wanted to know. Daniel explained that her total revenue was made up of two parts. The biggest part was her total sales, S. The other part was the sales tax she had collected, $(0.06)(S)$. Adding those two parts together makes the total revenue. So, Daniel's equation looked like this:

$$S + 0.06S = 1{,}367{,}250$$

Next Daniel explained that the sales plus the 6% tax is the same as $1.06S$

$$1.06S = 1{,}367{,}250$$

To find S, Daniel divided both sides of the equation by 1.06.

$$\frac{1.06\,S}{1.06} = \frac{1{,}367{,}250}{1.06}$$

$$S = \$1{,}367{,}250 \div 1.06 = \$1{,}289{,}858$$

Recall that Ellen was going to write a check to the McFadden Chicken Breast Co. for 10% of her total revenue, because she didn't know what her total sales were. However, now she does know!

12. Based on 10% of her total sales, find the amount of the check that Ellen wrote to the McFadden Chicken Breast Co.

13. Is the amount of the check Ellen wrote more or less than the amount of the check she was going to write?

14. How much was the difference?

Project idea:

Investigate on amazon.com or other online shopping sites to determine whether you pay sales tax on shipping and handling fees. Are shipping and handling fees included in the total cost before sales tax is calculated or are they added on after the sales tax is calculated?

Practice problems

1. Should the sales tax be included when calculating the tip amount or not? Explain your thinking.

2. Some restaurants include suggested tip amounts on the sales receipt. See example to the right. Does the Longhorn Restaurant include the sales tax when determining their suggested gratuities or not? Show your work.

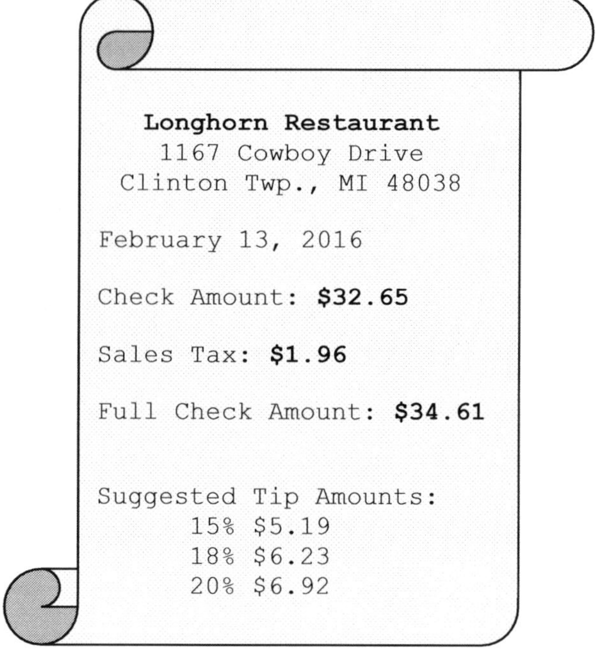
Longhorn Restaurant
1167 Cowboy Drive
Clinton Twp., MI 48038

February 13, 2016

Check Amount: **$32.65**

Sales Tax: **$1.96**

Full Check Amount: **$34.61**

Suggested Tip Amounts:
15% $5.19
18% $6.23
20% $6.92

Figure 2: Longhorn sales receipt

In the water bill below, the bar graph indicates the usage in gallons each month. Answer the following questions using the data from the bar graph below. The chart covers 12 months of bills.

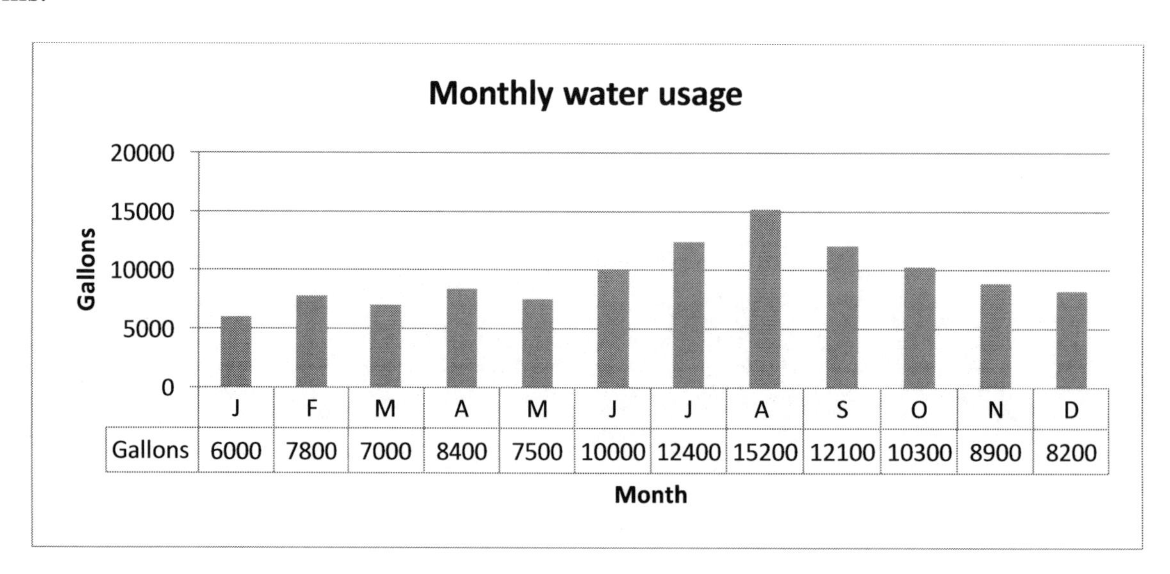

	J	F	M	A	M	J	J	A	S	O	N	D
Gallons	6000	7800	7000	8400	7500	10000	12400	15200	12100	10300	8900	8200

Figure 3: Monthly water usage

3. What month does the O represent?

4. What letters can be confusing? Why do you think only one letter was used to label each month?

5. By how many gallons did the water usage change from May to June? Find the percentage change from May to June.

6. By how many gallons did the water usage change from July to August? Find the percentage change from July to August.

7. Which increase in water usage was greater, between May and June or between July August? Which percentage change was greatest? Explain why the two answers are not the same.

8. By how many gallons did the water usage change from August to September? Find the percentage change from August to September.

9. Which change in water usage was greater, between July and August or between August and September? Which percentage change was greatest? Explain why the two answers are not the same.

10. What pattern can you observe about water usage in the months before and after August?

11. What value do you see in having a bar chart and not just a data table?

12. Coyotes in a local neighborhood have decreased the cat population by 40%. After local authorities capture the coyotes to release them in the wild, the cat population grows by 40%. Complete Table 3 for different amounts of the starting population of cats. Is the local neighborhood cat population restored? Why or why not?

Original cat population	Cat population after 40% decrease	Cat population after 40% increase	Percentage of cat population restored
200 Cats			
100 Cats	60 Cats	84 Cats	84%
50 Cats			
25 Cats			

Table 3: Cat population

13. While training for a marathon, Teresa increased her mileage per week by 10%. If she started with 40 miles per week and continued for 2 weeks before suffering shin splints and was forced to decrease her mileage by 10% per week for 2 weeks, is she back to her original mileage? Explain your reasoning.

Activity 7: Teachers' guide

Thinking through a lesson protocol

Standards:

6.RP.A.3.C: Find a percent of a quantity as a rate per 100 (e.g., 30% of a quantity means 30/100 times the quantity); solve problems involving finding the whole, given a part and the percent.

7.RP.A.3: Use proportional relationships to solve multistep ratio and percent problems.

Mathematical Practices:

MP1: Make sense of problems and persevere in solving them.

MP2: Reason abstractly and quantitatively.

MP3: Construct viable arguments and critique the reasoning of others.

Setting up the problem - Launch	
Selecting tasks/goal setting	(10 minutes) Ask students to discuss whether increasing a given amount by 50% and then decreasing the result by 50% gets you back to the original amount.
Questions	For example: Does increasing 100 by 50% and then decreasing the result by 50% get you back to 100? Why or why not?

Monitoring student work – Explore		
Strategies and misconceptions-Anticipating	**Who - Selecting and sequencing**	**Questions and statements - Monitoring**
(20 minutes) Have students read the McFadden Restaurant problem and answer questions #1-2.		Have students restate the Restaurant problem in their own words and discuss the missing $1,000.
After Question #2, be sure that the students understand to find percent increase/decrease you divide the change by the original amount.		(Example) Your allowance increased from $10/week to $12 per week. By what percentage did your allowance increase?
(30 minutes) Between Questions #2 and #3, the teacher needs to walk the students through the text and work out the mathematics together.		Have students highlight the associative law. Ask students what they notice about increasing and decreasing by the same amount.
(15 minutes) Problems #3 - #5 have students complete the table using a calculator and check answers with their partner.		
(15 minutes) Problems #6 - #10 have students read and answer with a partner. Discuss.		Point out where the 1.06 comes from and the differences with or without the sales tax included.

Monitoring individual student work - Explore		
Strategies and misconceptions - Anticipating	**Who - Selecting and sequencing**	**Questions and statements - Monitoring**
For off-task students or for students that seem to be self-conscious about you listening to them share.		I am just listening or looking to find out how you are working on the problem. This helps me think about what we will do later.
For students that appear to be stuck. Also for when you are having a difficult time understanding their strategies.		Can you tell me a little about your reading? Could you describe the relationship between percents and coins/dollars? How would you describe the problem in your own words? What facts do you have? Could you try it with simpler numbers? Fewer numbers?
For students that want to ask you questions, these are ways to uncover their thinking and judge to what extent you want to respond.		Tell me what you've thought about so far. What do you know? Why are you interested in more information about that? Let me say a little about that part. Tell me what you've thought about so far. What do you know?

Managing the discussion – Summarize	
Parts of discussion - Connecting	**Questions and statements - Connecting**
Launching the discussion: Select the problems in questions #6-14 that students are struggling with or you wish to share out.	Will team 1 start us off by sharing one way of working on this problem? Please raise your hand when you are ready to share your solution. What did you do first when you were working on this problem? Let's start by clearing up a few things about the problem. Let's list some key parts in this problem. What was unclear in the problem?
Eliciting and uncovering student strategies	Joe would you be willing to start us off? What have you found so far? Can you repeat that? Can you explain how you got that answer? How do you know? Walk us through your steps. Where did you begin? Can you show us?
Focusing on mathematical ideas	Can you explain why this is true? Does this method always work? How is Bob's method similar to Kelly's method? What do all the solutions have in common? What would happen if I changed the numbers to _____?
Encouraging interactions	Do you agree or disagree with Kahlil's idea? What do others think? Would someone be willing to repeat what Tom just said? Would anyone be willing to add on to what Sue just said?
Concluding the discussion	Can anyone tell me some of the big ideas that we learned today? How would you explain what we learned today to a 5th grader? Some of the key points from our discussion today are . . . Tomorrow we will continue our exploration of ______ beginning with the idea from today that __________.
Post lesson notes	You may wish to assign the practice problems that you feel would benefit the students.

Solutions to text questions

1. How much did the restaurant sales increase from February to March?

   $99,000 - $90,000 = $9,000

2. What were the restaurant's sales in March?

   $99,000

3. Fill in the correct values for the last column in the table.

Percent change	I = Increase	D = Decrease	Final value Multiply I×D
5%	1.05	0.95	0.998
10%	1.10	0.90	*0.990*
25%	1.25	0.75	*0.938*
33%	1.33	1.67	*0.891*
50%	1.50	0.50	*0.750*

Table 2: The result of equal percentage increases and decreases

4. As the percent change increases, what happens to the value of I × D?

   The final value is decreasing.

5. With a 50% increase and a 50% decrease, what percent of the original number is the final number?

   $0.750 \times 100\% = 75\%$

6. In January, what was the average sales per day?

   $$\frac{\$100{,}000}{31} = \$3{,}225.81$$

7. What was the average sales per day in February?

   $$\frac{\$90{,}000}{28} = \$3{,}214.29$$

8. Do you think Ellen and Daniel should worry about the difference in the average daily sales in February compared to January?

   $3225.81 – $3214.29 = $11.52 which is less than 1% of the sales for the days in January. No, I would not be worried.

9. Find the average sales per day in March.

$$\frac{\$99{,}000}{31} = \$3{,}193.55$$

10. Should Ellen and Daniel think that the restaurant's sales in January, February, and March were up-and-down, or pretty much the same?

The difference between the January, February and March sales is around 1%. I would say sales are pretty much the same.

11. What was the amount of the check Ellen was going to write?

10% of $1,367,250 is $136,725.00

12. Based on 10% of her total sales, find the amount of the check that Ellen wrote to the McFadden Chicken Breast Co.

10% of $1,289,858 is $128,985.80

13. Is the amount of the check Ellen wrote more or less than the amount of the check she was going to write?

$128,985.80 is less than $136,725.00

14. How much was the difference?

$136,725.00 – $128,985.80 = $7,739.20

Solutions to practice problems

1. In determining what amount to tip a server at a restaurant, should the sales tax be included when calculating the tip amount or not? Explain your thinking.

   Answers will vary.

   From LATimes: Glendale resident Lee Lanselle ate breakfast the other day at the Hill Street Cafe in La Cañada Flintridge. As he waited for his credit card receipt, he worked out the tip in his head.

   The receipt arrived and Lanselle was surprised that his estimate of a 15% tip was less than the "suggested gratuity" printed on the form. A closer look revealed that the recommended tip on the receipt included the full amount of the meal, including taxes.

   "This isn't the first time I've seen something like this," Lanselle, 56, told me. "More and more I see 'gratuity guidelines' or 'suggested gratuity' on the receipt. But this was the first time I stopped to think about it and realized that they were calculating the tip for the gross amount, with taxes."

   He said he knows this is a relatively minor matter, but doesn't this fly in the face of tipping conventions? "Aren't they supposed to calculate the tip before taxes?"

   The white-gloved etiquette mavens at the Emily Post Institute in Vermont say that restaurant tips should be 15% to 20% of the bill's pretax total.

   The Etiquette Scholar website agrees that you should "tip on the pretax amount of the bill, not on the total."

   Most participants in an online discussion on tipping at the Zagat website came down on the side of pretax gratuities, although one person allowed that "if service is good, I tip on the total because a server's job is hard work."

   Another person commented that "almost all establishments, if they include recommended tip percentages on the bill, include the tax as part of the tip percentage.

2. Some restaurants include suggested tip amounts on the sales receipt. See example below. Does the Longhorn Restaurant include the sales tax in their suggested gratuities or not? Show your work.

   \$5.19 is what % of \$32.65? 5.19 / 32.65 = 0.1589 or 16% while 5.19 / 34.61 = 0.1499 or 15%

   Also \$6.23 / 32.65 = 0.1908 or 19% or 6.23/34.61 = 0.1800 or 18% so it appears the Longhorn Restaurant includes sales tax in their suggested tip amount.

In the water bill below, the bar graph indicates the usage in gallons each month. Answer the following questions using the data from the bar graph below. The chart covers 12 months of bills.

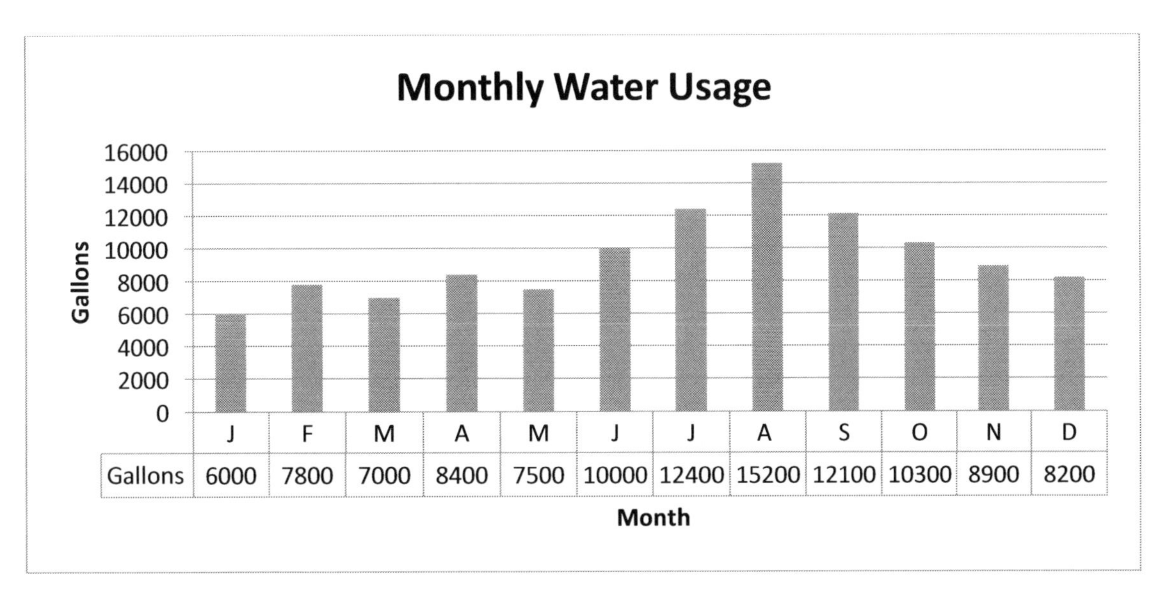

	J	F	M	A	M	J	J	A	S	O	N	D
Gallons	6000	7800	7000	8400	7500	10000	12400	15200	12100	10300	8900	8200

3. What month does the F represent?

   February

4. What letters can be confusing? Why do you think only one letter was used to label each month?

   The Js, Ms, and As might be confusing. One letter was possibly used to save space on the axis to match the width of the bars.

5. By how many gallons did the water usage change from May to June? Find the percentage change from May to June.

   The water usage increased by 2,500 gallons from May to June, 10,000 – 7,500 = 2,500. 2500 out of 7500 is a 33 1/3% increase, 2500 / 7500 = 0.333333 or 33 1/3%.

6. By how many gallons did the water usage change from July to August? Find the percentage change from July to August.

   The water usage increased by 2,800 gallons from July to August, 15,200 – 12,400 = 2,800. 2800 out of 12,400 is approximately a 22.6% increase, 2800 / 12400 = 0.225806 or 22.6%.

7. Which increase in water usage was greater, between May and June or between July and August? Which percentage change was greatest? Explain why the two answers are not the same.

There was a greater increase in gallons of water used between July and August, 2800 is greater than 2500. There was a greater percentage increase in water usage between May and June, 1/3% is greater than 22.6%. The two answers are not the same because the base number of gallons of water used is different in each case. The July to August percentage increase was on base of 12,400.

8. By how many gallons did the water usage change from August to September? Find the percentage change from August to September.

   The water usage decreased by 3,100 gallons from August to September, 12,100 – 15,200 = -3,100. 3,100 out of 15,200 is a 20.4% decrease, 3100 / 15200 = 0.203947 or 20.4%.

9. Which change in water usage was greater, between July and August or between August and September? Which percentage change was greatest? Explain why the two answers are not the same.

   There was a greater change in gallons of water used between August and September, 3100 is greater than 2800. There was a greater percentage change in water usage between July and August, 22.6% is greater than 20.4%. The two answers are not the same because the base number of gallons of water used is different in each case.

10. What pattern can you observe about water usage in the months before and after August?

    The months leading up to August, especially May, June, and July show an increase in water usage each month. The months following August show decreases each month in water usage.

11. What value do you see in having a bar chart and not just a data table?

    It is easy to see a change in water usage by comparing bars, also increases and decreases in water usage are easy to see based on height of bars.

12. Coyotes in a local neighborhood have decreased the cat population by 40%. After local authorities capture the coyotes to release them in the wild, the cat population grows by 40%. Complete Table 3 for different amounts of the starting population of cats. Is the local neighborhood cat population restored? Why or why not?

Original cat population	Cat population after 40% decrease	Cat population after 40% increase	Percentage of cat population restored
200 Cats	*120 Cats*	*168 Cats*	*84%*
100 Cats	60 Cats	84 Cats	84%
50 Cats	*30 Cats*	*42 Cats*	*84%*
25 Cats	*15 Cats*	*21 Cats*	*84%*

Table 3: Cat population

13. While training for a marathon, Teresa increased her mileage per week by 10%. If she started with 40 miles per week and continued for 2 weeks before suffering shin splints and was forced to decrease her mileage by 10% per week for 2 weeks, is she back to her original mileage? Explain your reasoning.

    Solution 1: No, because the % increase and the % decrease are the same over the same time span, therefore the final amount will always be less.

    Solution 2:

Week	*Start of week*	*Amount of 10% change*	*End of week*
1	*40 Miles*	*10 % of 40 = 4 (increase)*	*44 Miles*
2	*44 Miles*	*10 % of 44 = 4.4 (increase)*	*48.4 Miles*
injury			
3	*48.4 Miles*	*10% of 48.4 = 4.84 (decrease)*	*43.56 Miles*
4	*43.56 Miles*	*10% of 43.56 = 4.356 (decrease)*	*39.204 Miles*

Activity 8:

Managing Store Space

Mathematical goals

The student will explore how businesses use data from percentages to make management decisions. These decisions may involve advertising and marketing, how products are displayed and sold, as well as how many are produced.

The student will:

- Read a scenario and use data presented in tables
- Calculate different percentages
- Calculate averages of two numbers
- Use data to make a decision
- Work with percentages in a meaningful context familiar to students

Before the lesson (5-10 minutes)

Put the paper and pencil down and practice some mental mathematics.

- What is 25% of 800?
- There are 480 students in the school and 50% are boys. Of the boys, 40% are on the honor roll. How many boys are on the honor roll?
- Carol raises Gerbera Daisies in her greenhouse. Of the 200 plants she currently has, 20% are yellow, 75% are pink, and the remaining are red. How many of each color flower, does Carol have?

Managing by percentages – Two alternatives

All businesses gather data about their customers. They use the data to make decisions. One common statistic is the percentage of male and female customers. Stores use this percentage to decide how much space to give to different products. For example, in the clothing section of TJ Maxx and other stores, often more than 75% of the space is used for women's and girls' clothing. Another example is deciding how to divide their advertising money. What percentage should focus on women or on men?

Sometimes a business sees different percentages as a missed chance. Lego® was worried that in 2011 less than 10% of its customers were girls. They saw this low percentage as a chance to grow. They studied how boys and girls play with Lego® differently. They also studied what colors girls and boys prefer. They developed Lego® Friends and Lego® brand Disney Princess. They also used more colors. Lego®'s sales to young girls grew from $300 million in 2011 to $900 million in 2014[1].

---

[1] Adapted from http://fortune.com/2015/12/30/lego-friends-girls/

Bullseye Inc. – Men's and women's clothing

Bullseye Inc. has found that 70% of the clothing they sell is sold to women. They are opening a new store. The store will have 2500 square feet for clothing.

1. How many square feet should they use for women's clothing?

Bullseye also will advertise a lot when the store opens. They plan to spend $150,000 for magazine ads. There are a number of local magazines. Some are read mostly by men. Others are read mostly by women.

2. How much should they spend to advertise in men's magazines?

Toys Inc. – Action figures and dolls

Many stores use long shelves to display products. The available space is measured by the length of the shelves. Customers only see the items closest to the front.

Toys Inc. has found that their stores sell an average of $125,000 in action figures each year. They sell an average of $200,000 in dolls each year. They have three levels of shelves in the section for action figures and dolls. Each shelf is 60 feet long.

3. How many total feet of shelf space do they have for action figures and dolls?

4. How many feet of shelf space should they use for action figures?

Toys Inc. advertises on the Disney channel. The company has $20,000 this month for advertising action figures and dolls.

5. How much should they spend to advertise dolls?

6. How much should they spend to advertise action figures?

Farmer John's organic foods

Customers are buying more and more organic foods each year[2]. In 2014, 12% of all produce was organic. Customers are willing to pay more for organically grown fruits and vegetables. The percent increase in price is called the premium. Table 1 presents data for grapes and carrots. Organic grapes are 12% of the total grapes sold in this region. Customers are willing to pay 35% more for organic grapes. Organic carrots are 20% of the carrots sold in the region. The price premium is 18% more.

	Percent organic	Price premium
Grapes	12%	35%
Carrots	20%	18%

Table 1: Grapes and carrots

[2] For information about organic foods see https://ota.com/news/press-releases/18061

Farmer John's Market has set up 50 square feet of space for grapes. The basic price for non-organic grapes is $1.20 per pound.

7. How many square feet should be used for organic grapes?

8. What should Farmer John charge for organic grapes?

Farmer John's Market has set up 20 square feet of space for carrots. The basic price for non-organic carrots is $0.80 per pound.

9. How many square feet should be used for organic carrots?

10. What should Farmer John charge for organic carrots?

Project idea:

Ask students to visit their local grocery store and compare the cost of organic and non-organic foods. Find the percentage for the premium.

Investigate what produce you would want to buy as organic and what you don't feel would be as important to purchase organically? Explain your reasoning.

Practice problems

1. The Big Buy electronic store has an area designated for selling phones, tablets, and laptop computers. Big Buy currently uses 600 square feet to display these products. Based on marketing research, the area designated for the tablets is 25% of the total area, the phone area is 35%, and the laptop space allotted is 40% of the total area. How many square feet should be used for each display?

2. The principal at Carter Middle School has just purchased 500 lockers. The lockers are to be placed in four hallways: the north, the south, the east, and the west hallways. The hallways are not all the same size; there is a different amount of space in each hallway. The principal has decided to place 30% of the lockers in the north hallway, 25% in the east and west hallways, and 20% in the south hallway. How many lockers will there be in each of the hallways?

3. Customers are buying more and more organic foods each year. In 2014, 12% of all produce was organic. Customers are willing to pay more for organically grown fruits and vegetables. The percent increase in price is called the premium. Table 2 presents data for apples and strawberries. Organic apples are 15% of the total apples sold in this region. Customers are willing to pay 12% more for organic apples. Organic strawberries are 25% of the strawberries sold in the region. The price premium is 32% more.

	Percent organic	Price premium
Apples	15%	12%
Strawberries	25%	32%

Table 2: Apples and strawberries

Farmer John's Market has set up 60 square feet of space for apples. The basic price for non-organic apples is $2.00 per pound.

How many square feet should be used for organic apples?

4. What should Farmer John charge for organic apples?

Farmer John's Market has set up 20 square feet of space for strawberries. The basic price for non-organic strawberries is $1.20 per pound.

5. How many square feet should be used for strawberries?

6. What should Farmer John charge for organic strawberries?

Activity 8: Teachers' guide

Thinking through a lesson protocol

Standards:

6.RP.A.3.C: Find a percent of a quantity as a rate per 100 (e.g., 30% of a quantity means 30/100 times the quantity); solve problems involving finding the whole, given a part and the percent.

Mathematical Practices:

MP1: Make sense of problems and persevere in solving them.

MP2: Reason abstractly and quantitatively.

Setting up the problem - Launch	
Selecting tasks/goal setting	(5 minutes) Ask students if they have ever gone to a large department store that sells groceries and other items like clothing, household items, automotive accessories, etc. List some of the items they sell.
Questions	How do you think the manager of the department store decides how much space to assign to each type of item?

Monitoring student work – Explore		
Strategies and misconceptions- Anticipating	**Who - Selecting and sequencing**	**Questions and statements - Monitoring**
(10 Minutes) Read the first two paragraphs and ask students to paraphrase with a partner what they have just read.		
Continue to read the paragraph on Bullseye and answer questions #1-2. Share out whole class.		In answering Questions 1 & 2, have students justify their answers.
(15 minutes) Read Toys, Inc about action figures and have students work on solving questions #3-6. Share results.		
(20 minutes) Read the section on Farmer John's Organic Foods. Have students complete questions #7-8.		Ask students if they can think of other products that have premiums.
Brief discussion on premiums. Students complete questions #9-10. Check answers with partner.		

Monitoring individual student work - Explore		
Strategies and misconceptions - Anticipating	**Who - Selecting and sequencing**	**Questions and statements - Monitoring**
For off-task students or for students that seem to be self-conscious about you listening to them share.		I am just listening or looking to find out how you are working on the problem. This helps me think about what we will do later.
For students that appear to be stuck. Also for when you are having a difficult time understanding their strategies.		Can you tell me a little about your reading? How would you describe the problem in your own words? What facts do you have? Could you try it with simpler numbers?
For students that want to ask you questions, these are ways to uncover their thinking and judge to what extent you want to respond.		Tell me what you've thought about so far. What do you know? Why are you interested in more information about that? Let me say a little about that part. Tell me what you've thought about so far. What do you know?

Managing the discussion – Summarize	
Parts of discussion - Connecting	**Questions and statements - Connecting**
Launching the discussion: Select the problems in questions #1-10 that students are struggling with or you wish to share out.	Will team 1 start us off by sharing one way of working on this problem? Please raise your hand when you are ready to share your solution. What did you do first when you were working on this problem? Let's start by clearing up a few things about the problem. Let's list some key parts in this problem. What was unclear in the problem?
Eliciting and uncovering student strategies	Joe would you be willing to start us off? What have you found so far? Can you repeat that? Can you explain how you got that answer? How do you know? Walk us through your steps. Where did you begin? Can you show us?
Focusing on mathematical ideas	Can you explain why this is true? Does this method always work? How is Bob's method similar to Kelly's method? What do all the solutions have in common? What would happen if I changed the numbers to _____?
Encouraging interactions	Do you agree or disagree with Kahlil's idea? What do others think? Would someone be willing to repeat what Tom just said? Would anyone be willing to add on to what Sue just said?
Concluding the discussion	Can anyone tell me some of the big ideas that we learned today? How would you explain what we learned today to a 5th grader? Some of the key points from our discussion today are . . . Tomorrow we will continue our exploration of ______ beginning with the idea from today that __________.
Post lesson notes	You may wish to assign the practice problems that you feel would benefit the students.

Solutions to text questions

1. How many square feet should they use for women's clothing?

   70% of the 2500 square feet should be set aside for women's clothing
   0.7 × 2500 = ***1750 square feet***

2. How much should they spend to advertise in men's magazines?

   Since 70% of clothing is sold to women, 30% is sold to men and therefore 30% of the budget should be spent to advertise in men's magazines. 0.3 × $150,000 = ***$45,000***

3. How many total feet of shelf space do they have for action figures and dolls?

   3 × 60 feet = ***180 feet***

4. How many feet of shelf space should they use for action figures?

   The combined total of sales for action figures and dolls is $325,000. Action figures are $125,000 of the $325,000 in sales or roughly 38.5% of the sales. They should use 38.5% of the shelf space for action figures. 0.385 × 180 feet = ***69.3 feet***

5. How much should they spend to advertise dolls?

   Since 38.5% of sales are action figures, 61.5% of sales are dolls and 61.5% of dollars budgeted should be spent to advertise dolls. 0.615 × $20,000 = ***$12,308***

6. How much should they spend to advertise action figures?

   $20,000 – $12,308 = ***$7,692***

7. How many square feet should be used for organic grapes?

   12% of the 50 square feet 0.12 × 50 = ***6 square feet***

8. What should Farmer John charge for organic grapes?

   The premium for organic grapes is 35% above basic price for non-organic. Farmer John should charge $1.20 per pound + 0.35 × $1.20 per pound = ***$1.62 per pound***

9. How many square feet should be used for organic carrots?

   Since 20% of all carrots sold are organic, 20% of the available square feet should be used for organic carrots.

   0.2 × 20 square feet = ***4 square feet***

10. What should Farmer John charge for organic carrots?

    Farmer John should charge
    $0.80 per pound + 0.18 × $0.80 per pound = ***$0.95 per pound***

Solutions to practice problems

1. The Big Buy electronic store has an area designated for selling phones, tablets, and laptop computers. Big Buy currently uses 600 square feet to display these products. Based on marketing research, the area designated for the tablets is 25% of the total area, the phone area is 35%, and the laptop space allotted is 40% of the total area. How many square feet should be used for each display?

   Tablets: 25% of 600 is $0.25 \times 600 =$ ***150***

   Phone: 35% of 600 is $0.35 \times 600 =$ ***210***

   Laptop: 40% of 600 is $0.4 \times 600 =$ ***240***

2. The principal at Carter Middle School has just purchased 500 lockers. The lockers are to be placed in four hallways: the north, the south, the east, and the west hallways. The hallways are not all the same size; there is a different amount of space in each hallway. The principal has decided to place 30% of the lockers in the north hallway, 25% in the east and west hallways, and 20% in the south hallway. How many lockers will there be in each of the hallways?

   North Hallway: 30% of 500 is $0.3 \times 500 = =$ ***150***

   East Hallway: 25% of 500 is $0.25 \times 500 =$ ***125***

   West Hallway: 25% of 500 is $0.25 \times 500 =$ ***125***

   South Hallway: 20% of 500 is $0.2 \times 500 =$ ***100***

3. Customers are buying more and more organic foods each year3. In 2014, 12% of all produce was organic. Customers are willing to pay more for organically grown fruits and vegetables. The percent increase in price is called the premium. Table 2 presents data for apples and strawberries. Organic apples are 15% of the total apples sold in this region. Customers are willing to pay 12% more for organic apples. Organic strawberries are 25% of the strawberries sold in the region. The price premium is 32% more.

	Percent Organic	Price Premium
Apples	15%	12%
Strawberries	25%	32%

Table 2: Apples and strawberries

[3] For information about organic foods see https://ota.com/news/press-releases/18061

Farmer John's Market has set up 60 square feet of space for apples. The basic price for non-organic apples is $2.00 per pound.

How many square feet should be used for organic apples?

15% of 60 square feet is 0.15 × 60 = ***9 square feet***

4. What should Farmer John charge for organic apples?

   Farmer John should charge
   $2.00 per pound + 0.12 × $2.00 per pound = ***$2.24 per pound***

Farmer John's Market has set up 20 square feet of space for strawberries. The basic price for non-organic strawberries is $1.20 per pound.

5. How many square feet should be used for strawberries?

   25% of 20 square feet is 0.25 × 20 = ***5 square feet***

6. What should Farmer John charge for organic strawberries?

   Farmer John should charge
   $1.20 per pound + 0.32 × $1.20 per pound = ***$1.58 per pound***

Activity 9:
Multiple Flavor Ice Cream Sales

Mathematical goals

The student will use economic analysis and percentages to help decide the amount of each flavor of ice cream to purchase to resell.

The student will:

- Read and use data presented in table format
- Calculate percentages
- Determine costs, revenue and profit
- Solve meaningful mathematical problem and justify the answer.
- Transition to algebra to establish a policy.
- In this example the numbers do not work out perfectly. In the analysis a decision has to be made to buy a liter of ice cream even though there may be enough people to eat only half the ice cream. Whatever assumption is made should be justified.

Before the lesson (5-10 minutes)

Put the paper and pencil down and practice some mental mathematics.

Number talk possibilities:

Select two or three depending on student abilities.

- If 10 bite size candy bars cost $2.50, how much would one cost?
- If Kahlil buys 20 bottles of water for $4.00, how much does each bottle cost?
- Karen went to the dollar store for stickers. The cost of the stickers was 5 for $0.75. How much did she pay per sticker?
- Ken has scooped 20 ounces of ice cream out of the 80-ounce tub. What percentage of ice cream has he scooped?

Multiple flavor ice cream sales

During the summer, Charlie Haagen and his friend Dos sell ice cream cones on hot afternoons in the park. He expects to sell 300 ice cream cones this Sunday. He buys ice cream in one-liter tubs. Each one-liter tub costs $2.30. Charlie can fill ten cones from each tub. He buys large cones in packs of 5. Each 5-pack costs $0.60. He plans to sell each cone filled with ice cream for $1.00.

Charlie usually sells just four flavors of ice cream. He sold 1,200 ice cream cones the whole of last week. Charlie has kept careful records about sales of different flavors. Table 1 has his data.

Flavor	Number of cones	Percent
Vanilla	600	
Strawberry	300	
Chocolate Chip	180	
Mint	120	
Total	**1200**	

Table 1: Sales data for each flavor

1. Calculate the percent of customers who prefer each flavor.

2. When Mr. Haagen uses a whole one-liter tub, how much profit does he earn?

3. Mr. Hagen plans to sell 300 ice cream cones this Sunday. How many liters of each ice cream flavor do you recommend Mr. Haagen buy for this Sunday?

4. How much profit would Charlie make if he followed your suggestion?
   Clearly show all of your thinking.

Sometimes Charlie sells only part of a one-liter tub.

5. If he sells only eight ice cream cones of a particular flavor, does Charlie earn a profit?

6. If he sells only two ice cream cones of a particular flavor, does Charlie earn a profit?

Charlie has been learning some algebra in his math class. He thinks he can use algebra to find his break-even point. He breaks-even if the number of cones he sells brings in an amount equal to his costs.

7. Set up an equation to determine the break-even point for the number of cones sold. Be sure to state what the variable(s) in your equation represent(s).

8. Use the break-even point to set a buying policy when there are not enough customers to sell the entire one-liter tub.

Project idea:

Plan a fundraiser. Determine your pricing, cost, profit, revenue and what type of data you would need to collect.

Practice problems

The 6th Graders at Madison Middle School have an annual Bake Sale to raise money for the homeless. They collected data from their first 1500 sales over the past five years. They sell cupcakes by the dozen, cookies by the dozen, and cakes (10 inch size only). Data is provided in the Table 2.

	Sale price	Number sold	Percent of total sales	Total revenue	Cost	Total profit
Cupcakes (sold per dozen)	$4.80	500 dozen			$2.40 (per dozen)	
Cookies (sold per dozen)	$3.60	850 dozen			$1.20 (per dozen)	
Cakes	$8.00	150 cakes			$3.75	
Total						

Table 2: Cupcakes, cookies and cakes

1. Fill in the Percent of Sales column in Table 2 based on the data provided.
2. Fill in the Revenue Column in Table 2 for each item.
3. Fill in the Profit Column in Table 2.
4. Calculate the Total of Number Sold, Percent of Sales, and Profit.
5. Mrs. Jordan comes to the Bake Sale and asks if she can purchase 3 cupcakes. She wants to pay just one-quarter of the price per dozen. If they sell her the three cupcakes, but are not able to sell the other nine, will they still make a profit? Explain your thinking.
6. The 6th graders are considering selling individual cupcakes even though they can only buy cupcakes a dozen at a time. They will divide the sale price by 12. How many cupcakes from the dozen do they need to sell in order to cover the whole cost of the dozen they bought? (Covering the total cost is called breaking even.)

7. If the 6th graders want to break even on the sale of just three cupcakes, how much would they have to charge for each one?
8. Blair comes to the Bake Sale and asks if she can purchase 6 cookies for half the price of a dozen cookies. If they sell her the six cookies, but are not able to sell the other six, will they still make a profit? Explain your thinking.

9. The 6th graders are considering selling individual cookies for one-twelfth the price of a dozen. How many cookies from the dozen do they need to sell in order to break even?

10. At the end of their current bake sale, there are 6 cakes leftover. The students are trying to decide if they could discount them and still make a profit. Their advisor, Mrs. Larson indicated that she would like them to make a total profit of $1.50 from selling all of the remaining six cakes. How much should the students charge for each of the remaining six cakes?

11. Managers often want to compare the profitability of different products they sell. One measure that is used is the percent profit margin. This is calculated by dividing the profit per unit by the cost per unit. It is then converted to a percentage. For example, an item that costs $2.00 and is sold for $2.60 has a profit of $0.60 per unit. The percent profit margin is

$$\left(\frac{\$0.60}{\$2.00}\right)\times 100\% = 30\%$$

Which product, cupcakes by the dozen, cookies by the dozen, or individual cakes provides the highest percent profit margin?

Activity 9: Teachers' guide

Thinking through a lesson protocol

Standards:

6.RP.A.2: Understand the concept of a unit rate a/b associated with a ratio a:b with $b \neq 0$, and use rate language in the context of a ratio relationship.

6.RP.A.3.B: Solve unit rate problems including those involving unit pricing and constant speed.

6.RP.A.3.C: Find a percent of a quantity as a rate per 100 (e.g., 30% of a quantity means 30/100 times the quantity); solve problems involving finding the whole, given a part and the percent.

7.RP.A.3: Use proportional relationships to solve multistep ratio and percent problems.

Mathematical Practices:

MP1: Make sense of problems and persevere in solving them.

MP2: Reason abstractly and quantitatively.

MP3: Construct viable arguments and critique the reasoning of others.

MP6: Attend to precision.

Setting up the problem - Launch	
Selecting tasks/goal setting	(5 minutes) Briefly discuss or ask students in a whole group setting if they have ever been involved in a fundraiser? Does your school do any fundraisers?
Questions	What kind of information was recorded or needed to be collected?

<table>
<tr><th colspan="3">Monitoring student work – Explore</th></tr>
<tr><th>Strategies and misconceptions- Anticipating</th><th>Who - Selecting and sequencing</th><th>Questions and statements - Monitoring</th></tr>
<tr><td>(15 Minutes) Read the text and answer questions #1-4. Have students ready to share their solution to #4.</td><td></td><td rowspan="4">The teacher may want to gather several "number" equations before going to the general "variable" equation. (Refer back to the number equation in 5 & 6).</td></tr>
<tr><td>(10 Minutes) Share possible solutions for #4 with whole class.</td><td></td></tr>
<tr><td>(15 minutes) Discuss with the class when a business "breaks even". Have students work problems 5 & 6 with a partner.
Share results.</td><td></td></tr>
<tr><td>(20 minutes)
With a partner or in small groups, have students try to write the equations and answer 7 & 8.</td><td></td></tr>
</table>

Monitoring individual student work - Explore		
Strategies and misconceptions - Anticipating	**Who - Selecting and sequencing**	**Questions and statements - Monitoring**
For off-task students or for students that seem to be self-conscious about you listening to them share.		I am just listening or looking to find out how you are working on the problem. This helps me think about what we will do later.
For students that appear to be stuck. Also for when you are having a difficult time understanding their strategies.		Can you tell me a little about your reading? How would you describe the problem in your own words? What facts do you have? Could you try it with simpler numbers?
For students that want to ask you questions, these are ways to uncover their thinking and judge to what extent you want to respond.		Why are you interested in more information about that? Let me say a little about that part. Tell me what you've thought about so far. What do you know?

Managing the discussion – Summarize	
Parts of discussion - Connecting	**Questions and statements - Connecting**
Launching the discussion: Select the problems in questions #5-8 that students are struggling with or you wish to share out.	Will team 1 start us off by sharing one way of working on this problem? Please raise your hand when you are ready to share your solution. What did you do first when you were working on this problem? Let's start by clearing up a few things about the problem. Let's list some key parts in this problem. What was unclear in the problem?
Eliciting and uncovering student strategies	Joe would you be willing to start us off? What have you found so far? Can you repeat that? Can you explain how you got that answer? How do you know? Walk us through your steps. Where did you begin? Can you show us?
Focusing on mathematical ideas	Can you explain why this is true? Does this method always work? How is Bob's method similar to Kelly's method? What do all the solutions have in common? What would happen if I changed the numbers to _____?
Encouraging interactions	Do you agree or disagree with Kahlil's idea? What do others think? Would someone be willing to repeat what Tom just said? Would anyone be willing to add on to what Sue just said?
Concluding the discussion	Can anyone tell me some of the big ideas that we learned today? How would you explain what we learned today to a 5th grader? Some of the key points from our discussion today are . . . Tomorrow we will continue our exploration of ______ beginning with the idea from today that __________.
Post lesson notes	You may wish to assign the practice problems that you feel would benefit the students.

Solutions to text questions

Flavor	Number of cones	X Percent	Ice cream cones X% of 300	Liters	Purchase
Vanilla	600	*50%*	*150*	*15*	*15*
Strawberry	300	*25%*	*75*	*7.5*	*7 or 8?*
Choc. Chip	180	*15%*	*45*	*4.5*	*4 or 5?*
Mint	120	*10%*	*30*	*3*	*30*
Total	**1200**		**300**	**30**	

Table 1: Sales data for each flavor with *Calculations*

1. Calculate the percent of customers who prefer each flavor?

   See Table 1

2. When Mr. Haagen uses a whole one-liter tub, how much profit does he earn?

   One liter can fill ten cones.
   Cost for ten cones = \$2.30 ice cream + 2(\$0.60) cones = \$3.50
   Revenue 10 × \$1.00 = \$10.00
   Profit = Revenue – Cost = \$10.00 – \$3.50 = \$6.50 for every ten ice cream cones

3. Mr. Hagen plans to sell 300 ice cream cones this Sunday. How many liters of each ice cream flavor do you recommend Mr. Haagen buy for this Sunday?

   Multiply the percent preference by 300 people. See table
   Strawberry and chocolate pose a problem. The results suggest half liter purchases but this is not allowed.
   Options:
   a) Order only 7 and 4 and miss out on 10 customers
   b) Order 8 strawberry and 4 chocolate chip liters. Assume you may be able to convince a few chocolate chip lovers to buy strawberry instead.
   c) Order 7 strawberry and 5 chocolate chip liters. Assume you may be able to convince a few strawberry lovers to buy chocolate chip instead.
   d) Order 8 strawberry and 5 chocolate chip knowing there will be left over ice cream.

4. How much profit would Charlie make if he followed your suggestion?

   Let's assume he buys 8 strawberry and 4 chocolate chip and sells everything.
   He sells 30 liters and makes \$6.50 profit each = 30 × \$6.50 = \$195

5. If customers buy just eight ice cream scoops of a particular flavor, does Charlie earn a profit?

   A liter costs $2.30 and two packs of cones cost $1.20 for a total of $3.50. He sells 8 cones for a total of $8.00. His profit is $4.50

6. If customers buy just two ice cream scoops of a particular flavor, does Charlie earn a profit?

   A liter costs $2.30 and one pack of cones cost $0.60 for a total of $2.90. He sells 2 cones for a total of $2.00. His net loss is $0.90

7. Algebra: Set up an equation to determine the break-even point for the number of customers for which the revenue and costs are equal. Use this information to set a liter buying policy when there are not enough customers to eat the entire liter.

   Assume some number less than 5 will be the break-even point for x the number of cones sold.
   Cost $2.30 + 0.60 = $1x.

8. Use the break-even point to set a buying policy when there are not enough customers to sell the entire one-liter tub.

   Cost $2.30 + 0.60 = $1x. Thus x=2.9 liters. This is not an option. If we round up to 3, he earns a profit. With 3 customers, the profit would be $0.10.
   Policy: if 3 or more customers want a flavor buy a liter of that ice cream.

Solutions to practice problems

The 6th Graders at Madison Middle School have an annual Bake Sale to raise money for the homeless. They collected data from their first 1500 sales over the past five years. They sell cupcakes by the dozen, cookies by the dozen, and cakes (10 inch size only). Data is provided in the table below.

	Sale price	Number sold	Percent of total sales	Total revenue	Cost	Total profit
Cupcakes (sold per dozen)	$4.80	500 dozen	*33 ⅓%*	*$2400*	$2.40 (per dozen)	*$1200*
Cookies (sold per dozen)	$3.60	850 dozen	*56 ⅔%*	*$3060*	$1.20 (per dozen)	*$2040*
Cakes	$8.00	150 cakes	*10%*	*$1200*	$3.75	*$637.50*
Total		*1500*	*100%*			*$3877.50*

Table 2: Cupcakes, cookies and cakes

1. Fill in the Percent of Sales column based on the data provided.

   See Table 2.

2. Fill in the Revenue Column in the table above for each item.

   See Table 2.

3. Fill in the Profit Column above.

   See Table 2.

4. Complete the shaded TOTAL columns for Number Sold, Percent of Sales, and Profit.

   See Table 2.

5. Mrs. Jordan comes to the Bake Sale and asks if she can purchase 3 cupcakes. She wants to pay just one-quarter of the price per dozen. If they sell her the three cupcakes, but are not able to sell the other nine, will they still make a profit? Explain your thinking.

   The three cupcakes will cost Mrs. Jordan $1.20 ($0.40/cupcake). To make the entire 12 cupcake (dozen) batch, the cost is $2.40. This would indicate a loss of $1.20,

   $2.40 – $1.20=$1.20.

6. The 6th graders are considering selling individual cupcakes even though they can only buy cupcakes a dozen at a time. They will divide the sale price by 12. How many cupcakes from the dozen do they need to sell in order to cover the whole cost of the dozen they bought? (Covering the total cost is called breaking even.)

   They will need to sell at least 6 cupcakes ($0.40/cupcake) for $2.40 in order for them to break even with the cost of the batch of cupcakes which is $2.40.

7. If the 6th graders want to break even on the sale of just three cupcakes, how much would they have to charge for each one?

   In order to break even with the cost of the batch of cupcakes which is $2.40, to sell three would require a charge of $0.80, 3 × $0.80 = $2.40.

8. Blair comes to the Bake Sale and asks if she can purchase 6 cookies for half the price of a dozen cookies. If they sell her the six cookies, but are not able to sell the other six, will they still make a profit? Explain your thinking.

   Each cookie sells for $0.30. If they sell 6 cookies out of the dozen, the revenue would be $1.80 which is $0.60 more than the cost of making the dozen. They would make a profit.

9. The 6th graders are considering selling individual cookies for one-twelfth the price of a dozen. How many cookies from the dozen do they need to sell in order to break even?

   Since each cookie sells for $0.30 and the cost for making a dozen is $1.20 they would need to sell 4 for break even. 4 × $0.30 = $1.20 OR $1.20 – (n / 12)($0.30) > 0

10. At the end of their current bake sale, there are 6 cakes leftover. The students are trying to decide if they could discount them and still make a profit. Their advisor, Mrs. Larson indicated that she would like them to make a total profit of $1.50 from selling all of the remaining six cakes. How much should the students charge for each of the remaining six cakes?

    To distribute the profit evenly, $1.50 / 6 = $0.25, each cake should sell for $3.75 + $0.25 or $4.00. 50% of the original selling price.

11. Managers often want to compare the profitability of different products they sell. One measure that is used is the percent profit margin. This is calculated by dividing the profit per unit by the cost per unit. It is then converted to a percentage. For example, an item that costs $2.00 and is sold for $2.60 has a profit of $0.60 per unit. The percent profit margin is

    $$\left(\frac{0.6}{2.00}\right)\times 100\% = 30\%$$

    Which product, cupcakes by the dozen, cookies by the dozen, or individual cakes provides the highest percent profit margin?

The profit for cupcakes is $2.40 and so is the cost. The percent profit is 100%. The profit on cookies is also $2.40 but the cost is only $1.20. The percent profit is 200% which the largest. Cakes cost $3.75 and earn $4.25 profit. The percent profit is 113%. Cookies have the highest percent profit relative to their cost.

	Sale price	**Cost**	**Profit per unit**	**Percent profit = (profit /cost) × 100%**
Cupcakes (sold per dozen)	$4.80	$2.40 (per dozen)	*$2.40*	*(2.40 / 2.40) × 100% = 100%*
Cookies (sold per dozen)	$3.60	$1.20 (per dozen)	*$2.40*	*(0.40 / 1.20) × 100% = 200%*
Cakes	$8.00	$3.75	*$4.25*	*(4.25 / 3.75) × 100% = 113%*

Activity 10:
Ordering Hoodies
Multiple Percentages

Mathematical goals

The student will use market information to help decide the number of men's and women's sweatshirts to order by size and color.

The student will:

- Read and use data presented in table format.
- Read and use data presented in a pie chart.
- Calculate percentages.
- Use pairs of percentages, size and color, to make decisions.
- Solve a mathematical problem set in a meaningful context and justify the answer.
- Rounding numbers to the nearest integer.

In this example the numbers do not work out perfectly. It is not possible to order a fraction of a sweatshirt. In addition, the rounding up and rounding down must be done carefully to ensure the total is constant. Whatever assumption is made should be justified.

Before the lesson (5-10 minutes)

Put the paper and pencil down and practice some mental mathematics.

Number Talk Possibilities:

Select two or three depending on student abilities.

- What is 10% of 240?
- What is 20% of 240?
- What is 5% of 240?
- What is 35% of 240?
- What is 60% of 240?

Ordering hoodies - Multiple percentages

Clothing stores order the clothes they sell in different sizes. The person who orders must decide what percentage of sweatshirts should be 2XL, XL, L, M and S. This decision is based on the percentage of people who wear different sizes. For many things, stores also order different colors. What percentage of t-shirts should be white, blue, or pink? These color percentages will be different for men's and women's clothing. Knowing the percentage of people who prefer specific colors helps to make the decision. For example, the colors buyers want has a big impact on the colors of the cars at a car dealership. Across the globe, white is the most popular color. Over 20% of people choose white. More than 70% of customers worldwide choose white, black, silver or grey[1].

Hoodie Gallery, Inc.

Hoodie Gallery is a store that sells sweatshirts. They have a new sweatshirt with a super-star athlete's brand-name. They will order 100 men's and 100 women's sweatshirts of all sizes. Dawn Garb, the person who orders, went to a website to find the percentage of sizes for men and women. She found the information in Table 1 online[2].

[1] You can find information like this online at https://en.wikipedia.org/wiki/Car_colour_popularity.
[2] https://blog.adafruit.com/2011/10/04/most-common-sizes-for-shirts/Small

	Men's		Women's	
Size	Percent	# of shirts to order	Percent	# of shirts to order
Small	3%		19%	
Medium	13%		39%	
Large	37%		29%	
Xtra Large	34%		10%	
2XL	13%		3%	

Table 1: Estimated percentage of sizes for men's and women's sweatshirts

1. If Dawn uses Table 1 as a guide, how many of each size should she order? Fill in the table.

Dawn asked her data analyst, Celine, to make a graph from the table. Celine chose to make pie charts. The size of each slice of the pie represents the corresponding percentage. She also used a different shading for each of the different sizes. Celine used Excel to make Figures 1 and 2.

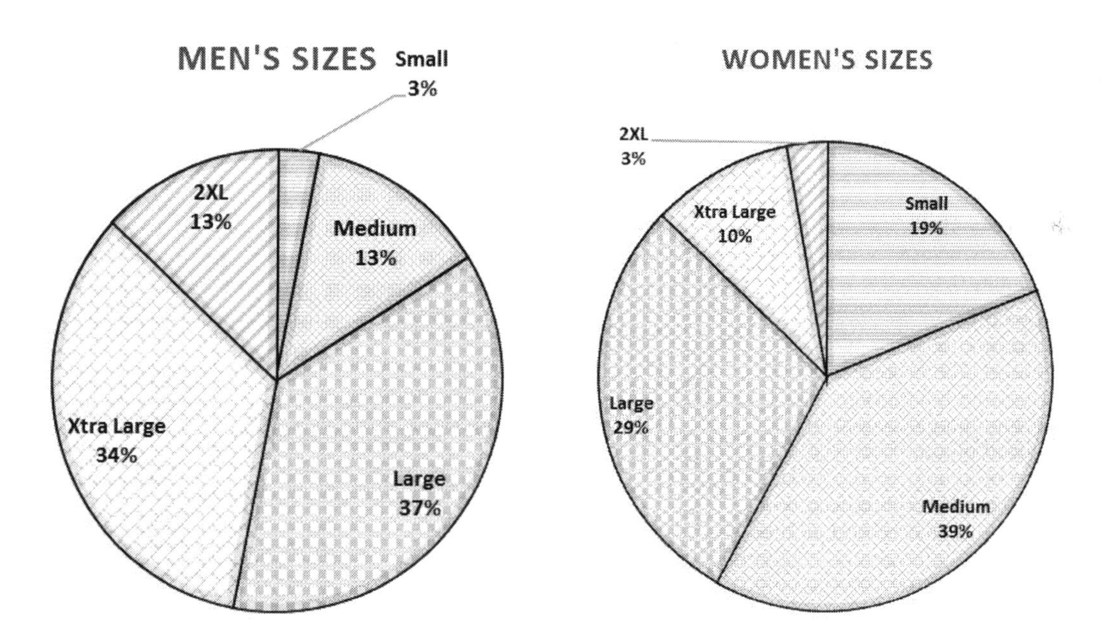

Figure 1: Pie chart of men's sizes

Figure 2: Pie chart of women's sizes

2. In the men's sizes, which slices of the pie were the same size?

3. How much larger is the Xtra Large pie slice as compared to the 2XL pie slice in men's clothing?

4. How much bigger is the Xtra Large slice than the 2XL slice in women's clothing?

5. How are the men's percentages different from the women's?

When she looked at the pie charts, Dawn decided not to order any size S for men nor 2XL for women.

6. Why do you think she made this decision?

Men's			Women's	
Size	**Percent**	**# of shirts to order**	**Percent**	**# of shirts to order**
Small		0	19%	
Medium	13%		39%	
Large	37%		29%	
Xtra Large	34%		10%	
2XL	13%			0

Table 2: Revised number of sizes for men's and women's sweatshirts

Dawn still plans to order 100 each of men's and women's sweatshirts. The remaining percentages now add up to 97%. She can use these percentages to order 97 men's and 97 women's sweatshirts.

7. What sizes should she order for the three remaining men's sweatshirts? Justify your answer. Record your revised order in Table 2.

8. What sizes should she order for the three remaining women's sweatshirts? Justify your answer. Record your revised order in Table 2.

The men's sweatshirts come in two colors: charcoal grey and navy blue. Women's colors are azure and fuchsia. Dawn asked Celine to gather data on which colors men and women prefer. Celine surveyed 120 men and 120 women. She put the survey results in Table 3.

Men			Women		
Color	**Survey**	**Percent**	**Color**	**Survey**	**Percent**
Charcoal Grey	90		**Azure**	70	
Navy Blue	30		**Fuchsia**	50	

Table 3: Men and women color preferences

9. Find the percentages for each color for men and women and write them in Table 3.

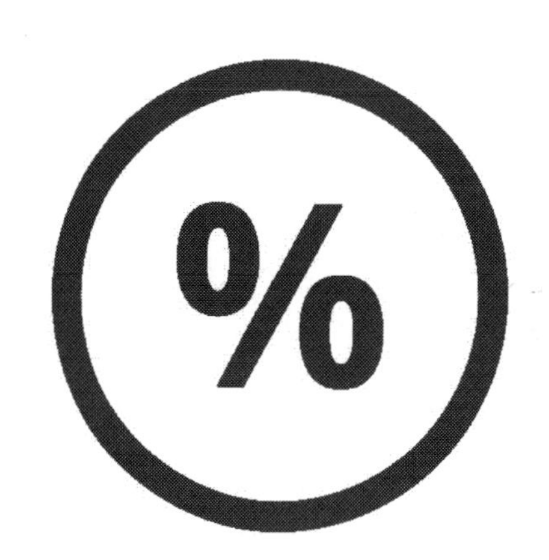

Dawn now had to decide how many of each color to order for each size. She began with the men's sweatshirts.

10. Record the men's order plan from Table 2 in the Total by Size column in Table 4. Also, take from Table 3 the percent of each color and record them above the Charcoal Grey and Navy Blue columns.

Size	**Number of men's sweatshirts**		
	Total by size	**Charcoal grey** Percent =	**Navy blue** Percent =
Medium			
Large			
Xtra Large			
2XL			
Totals			

Table 4: Number of men's sweatshirts to order by size and color.

11. Use the color percentages of men to find the number of sweatshirts of each size and color Dawn should order. Write this information in Table 4.

12. What was difficult about deciding the number of sweatshirts of each size and color? How did you handle the difficulty?

13. Record the women's order plan from Table 2 in the Total by Size column in Table 5. Also, record the percent of each color in the second row.

14. Repeat the above process for women's sweatshirts to determine how many of each color to order for each size. Record the answers in Table 5.

Size	Number of Women's Sweatshirts		
	Total by Size	Azure	Fuchsia
		Percent =	Percent =
Small			
Medium			
Large			
Xtra Large			
Totals			

Table 5: Number of women's sweatshirts to order by size and color.

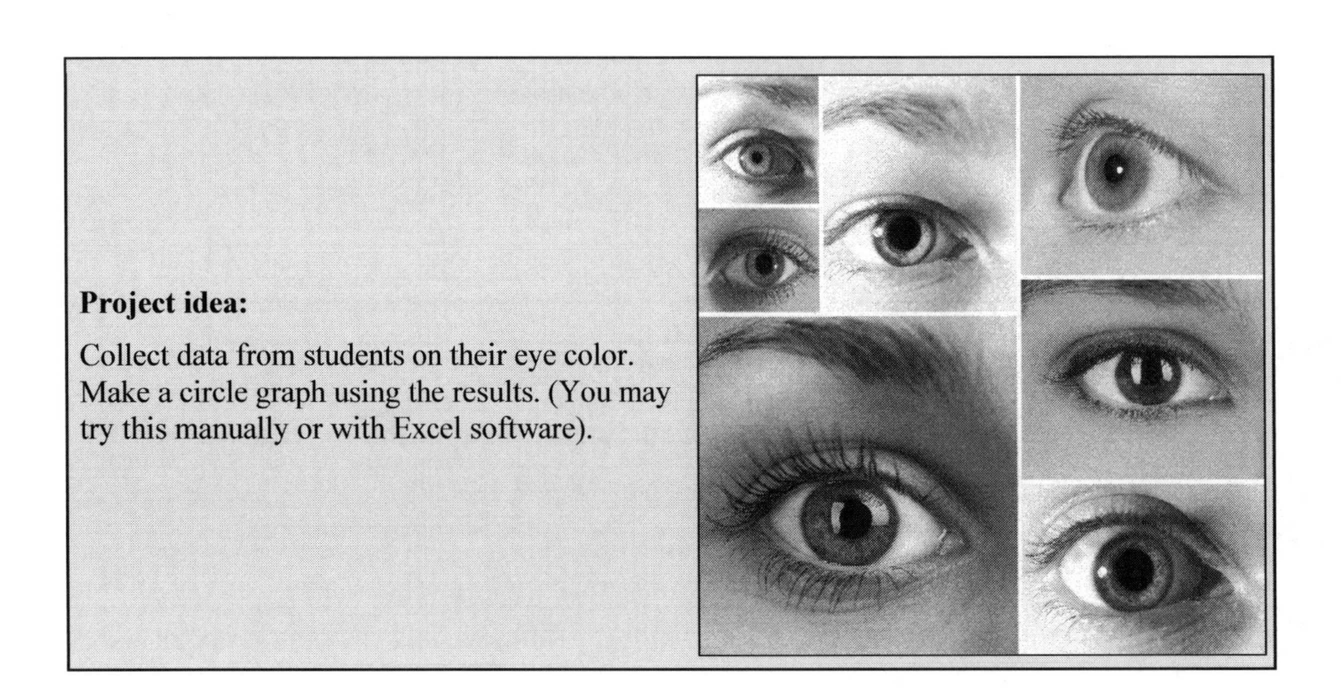

Project idea:

Collect data from students on their eye color. Make a circle graph using the results. (You may try this manually or with Excel software).

Practice problems

Nike hired Al Wright to ask the students at Covington High School to take a survey of their favorite tennis shoe. Students were required to identify themselves by gender, shoe size, and favorite color. The data below shows the results of 500 students who returned their surveys.

Men			Women		
Color	**# of men**	**Percent**	**Color**	**# of women**	**Percent**
Black	110		**Black**	25	
White	23		**White**	180	
Grey	35		**Green**	25	
Blue	17		**Blue**	60	
Orange	15		**Orange**	10	
Totals			**Totals**		

Table 6: Tennis shoe color data survey

1. How many men returned their survey according to the Tennis Shoe Color Data above?

2. How many women returned their survey according to the Tennis Shoe Color Data above?

3. Complete the Tennis shoe color data table by filling in the missing percentages for both men and women.

4. Write four statements that you can make based on the Shoe Color Data table above.

5. Bernice Barter is the school sales rep for Nike. She is planning on selling 80 pair of discount Nike tennis shoes to the boys in the school. How many of each color should she stock in the store at Covington High School? Fill in Table 7. Justify your reasoning.

6. She thinks she can sell 120 pair of discount Nike tennis shoes to the girls in the school. Justify her reasoning using the survey data in Table 6.

7. How many of each color women's shoe should she stock in the store at Covington High School? Fill in Table 7.

Men			Women		
Color	Percent	# of men	Color	Percent	# of women
Black			Black		
White			White		
Grey			Green		
Blue			Blue		
Orange			Orange		
Totals		80	Totals		120

Table 7: Tennis shoe color order

The survey also asked the students to report their shoe sizes. Table 8 contains the shoe size data. The data were tabulated and the percentages for each shoe size were calculated.

Men	Size	9 1/2	10	10 1/2	11	11 1/2	TOTAL
	Percent	8%	19%	40%	23%	10%	100%
Women	Size	7 1/2	8	8 1/2	9	9 1/2	
	Percent	12%	17%	45%	18%	8%	100%

Table 8: Tennis shoe size data survey

8. Write four statements that you can make based on Table 8.

When placing the order, Bernice knew she had to specify the sizes for each color. Take the men's color information in Table 7 and record it in the "Totals" column of Table 9.

9. Next use the information in Table 8 to help Bernice place her men's order.

		Totals	Men's shoe size				
			9 1/2	10	10 1/2	11	11 1/2
			8%	19%	40%	23%	10%
Color	Black						
	White						
	Grey						
	Blue						
	Orange						

Table 9: Men's tennis shoe color order by size

		Totals	Women's shoe size				
			7 1/2	8	8 1/2	9	9 1/2
			12%	17%	45%	18%	8%
Color	Black						
	White						
	Grey						
	Blue						
	Orange						

Table 10: Women's tennis shoe color order by size

Finally, take the women's color information in Table 7 and record it in the "Totals" column of Table 10.

10. Next use the information in Table 8 to help Bernice place her women's order for each color and size.

Activity 10: Teachers' guide

Thinking through a lesson protocol

Standards:

6.RP.A.3.C: Find a percent of a quantity as a rate per 100 (e.g., 30% of a quantity means 30/100 times the quantity); solve problems involving finding the whole, given a part and the percent.

7.RP.A.3: Use proportional relationships to solve multistep ratio and percent problems.

Mathematical Practices:

MP1: Make sense of problems and persevere in solving them.

MP2: Reason abstractly and quantitatively.

MP3: Construct viable arguments and critique the reasoning of others.

MP6: Attend to precision.

Setting up the problem - Launch	
Selecting tasks/goal setting	(5 minutes) What do you think is the most popular car color for boys? How about the most popular for girls?
Questions	If you owned a car dealership, what might be other factors to consider when ordering cars to sell?

Monitoring student work – Explore		
Strategies and misconceptions-Anticipating	**Who - Selecting and sequencing**	**Questions and statements - Monitoring**
(10 Minutes) Read the text and discuss and complete question #1. Have students compare their table with a neighbor.		
(10 Minutes) Continue reading about circle graphs and work problems # 2-6.		Depending on student's prior knowledge of a circle graph, discuss how the graph is made. Share out and discuss attributes of the circle graph.
(10 minutes) Continue with questions 7 & 8. Ask a few students to share their thinking.		
(20 minutes) Read text after question 8. Calculate and complete # 9-14.		

Monitoring individual student work - Explore		
Strategies and misconceptions - Anticipating	**Who - Selecting and sequencing**	**Questions and statements - Monitoring**
For off-task students or for students that seem to be self-conscious about you listening to them share.		I am just listening or looking to find out how you are working on the problem. This helps me think about what we will do later.
For students that appear to be stuck. Also for when you are having a difficult time understanding their strategies.		Can you tell me a little about your reading? How would you describe the problem in your own words? What facts do you have? Could you try it with simpler numbers?
For students that want to ask you questions, these are ways to uncover their thinking and judge to what extent you want to respond.		Why are you interested in more information about that? Let me say a little about that part. Tell me what you've thought about so far. What do you know?

Managing the discussion – Summarize	
Parts of discussion - Connecting	**Questions and statements - Connecting**
Launching the discussion: Select the problems in questions #9-14 that students are struggling with or you wish to share out.	Will team 1 start us off by sharing one way of working on this problem? Please raise your hand when you are ready to share your solution. What did you do first when you were working on this problem? Let's start by clearing up a few things about the problem. Let's list some key parts in this problem. What was unclear in the problem?
Eliciting and uncovering student strategies	Joe would you be willing to start us off? What have you found so far? Can you repeat that? Can you explain how you got that answer? How do you know? Walk us through your steps. Where did you begin? Can you show us?
Focusing on mathematical ideas	Can you explain why this is true? Does this method always work? How is Bob's method similar to Kelly's method? What do all the solutions have in common? What would happen if I changed the numbers to _____?
Encouraging interactions	Do you agree or disagree with Kahlil's idea? What do others think? Would someone be willing to repeat what Tom just said? Would anyone be willing to add on to what Sue just said?
Concluding the discussion	Can anyone tell me some of the big ideas that we learned today? How would you explain what we learned today to a 5th grader? Some of the key points from our discussion today are . . . Tomorrow we will continue our exploration of ______ beginning with the idea from today that __________.
Post lesson notes	You may wish to assign the practice problems that you feel would benefit the students.

Solutions to text questions

1. If Dawn uses this table as a guide, how many of each size should she order? Fill in the table.

Size	Men's		Women's	
	Percent	# of shirts to order	Percent	# of shirts to order
Small	3%	*3*	19%	*19*
Medium	13%	*13*	39%	*32*
Large	37%	*37*	29%	*29*
Xtra Large	34%	*34*	10%	*10*
2XL	13%	*13*	3%	*3*

Table 1: Estimated percentage of sizes for men's and women's sweatshirts

2. In the men's sizes, which slices of the pie were the same size?

   The slice representing Medium and the slice representing 2XL.

3. How much larger is the Xtra Large pie slice as compared to the 2XL pie slice in men's clothing?

   *34% - 11% = **23%** larger*

4. How much bigger is the Xtra Large slice than the 2XL slice in women's clothing?

   *10% - 3% = **7%** larger*

5. How are the men's percentages different from the women's?

   ***Answers will vary**. The men's sizes Xtra Large and L make up more than 50% of the sizes. The women's sizes Xtra Large and L are less than 50% of the sizes.*

6. Why do you think she made this decision?

   3% is very small

7. What sizes should she order for the three remaining men's sweatshirts? Justify your answer. Record your revised order in Table 2.

   ***Answer will vary.** Since the largest percentage is 37% and the next largest is 34%, she should order 2 more Large and 1 more Xtra Large. See Table 2.*

Size	Men's		Women's	
	Percent	# of shirts to order	Percent	# of shirts to order
Small		0	19%	*19*
Medium	13%	*13*	39%	*41*
Large	37%	*39*	29%	*30*
Xtra Large	34%	*35*	10%	*10*
2XL	13%	*13*		0

Table 2: Revised number of sizes for men's and women's sweatshirts

8. What sizes should she order for the three remaining women's sweatshirts? Justify your answer. Record your revised order in Table 2.

   Answers will vary. *Since the largest percentage is 39% and the next largest is 29%, she should order 2 more Medium and 1 more Large. See Table 2.*

9. Find the percentages for each color for men and women and write them in Table 3.

Color	Men		Women		
	Survey	Percent	Color	Survey	Percent
Charcoal Grey	90	*75%*	**Azure**	70	*58%*
Navy Blue	30	*25%*	**Fuchsia**	50	*42%*

Table 3: Men and women color preferences

10. Record the men's order plan from Table 2 in the Total by Size column in Table 4. Also, take from Table 3 the percent of each color and record them above the Charcoal Grey and Navy Blue columns.

Size	Number of men's sweatshirts		
	Total by size	Charcoal grey Percent = *75%*	Navy blue Percent = *25%*
Medium	*13*	*10*	*3*
Large	*39*	*29*	*10*
Xtra Large	*35*	*26*	*9*
2XL	*13*	*10*	*3*
Totals	*100*	*75*	*25*

Table 4: Number of men's sweatshirts to order by size and color

11. Use the color percentages of men to find the number of sweatshirts of each size and color Dawn should order. Write this information in Table 4.

    Answers will vary. *The answers in the table are rounded off. For example with medium,* $0.75 \times 13 = 9.75$ *and* $0.25 \times 13 = 3.25$*. These were rounded to 10 and 3 respectively.*

12. What was difficult about deciding the number of sweatshirts of each size and color?

    You can't order a fractional part of a sweatshirt.

    How did you handle the difficulty?

    If I ignored the fractional part I was not ordering less than 100 sweatshirts. If I always went up I was ordering more than 100 sweatshirts. I decided to round to the nearest whole number.

13. Record the women's order plan from Table 2 in the Total by Size column in Table 5. Also, record the percent of each color in the second row.

    See answers in Table 5 below.

14. Repeat the above process for women's sweatshirts to determine how many of each color to order for each size. Record the answers in Table 5.

    Answers will vary.

Size	Number of Women's Sweatshirts		
	Total by Size	Azure	Fuchsia
		Percent = *58%*	Percent = *42%*
Small	*19*	*11*	*8*
Medium	*41*	*24*	*17*
Large	*30*	*17*	*13*
Xtra Large	*10*	*6*	*4*
Totals	*100*	*58*	*42*

Table 5: Number of women's sweatshirts to order by size and color.

Solutions to practice problems

Men			Women		
Color	**# of men**	**Percent**	**Color**	**# of women**	**Percent**
Black	110	*55%*	**Black**	25	*8.3%*
White	23	*11.5%*	**White**	180	*60%*
Grey	35	*17.5%*	**Green**	25	*8.3%*
Blue	17	*8.5%*	**Blue**	60	*20%*
Orange	15	*7.5%*	**Orange**	10	*3.3%*
Totals	*200*	*100%*	**Totals**	*300*	*100%*

Table 6: Tennis shoe color data survey

1. How many men returned their survey according to Table 1?

   200 men returned their survey

2. How many women returned their survey according to Table 1?

   300 women returned their survey

3. Complete Table 1 by filling in the missing percentages for both men and women.

   See Table 1 above.

4. Write four statements that you can make based on the Shoe Color Data table above.

   Answers will vary.
   Example 1: The most popular shoe color for men's tennis shoes is black.
   Example 2: Grey shoe color for men is more popular than blue and orange combined.

5. Bernice Barter is the school sales rep for Nike. She is planning on selling 80 pair of discount Nike tennis shoes to the boys in the school. How many of each color should she stock in the store at Covington High School? Fill in Table 7. Justify your reasoning.

Multiply the percentages in Table 6 by 80 to obtain the number to order. These appear in Table 7 below. Some of the values will be non-integer. We rounded off the values to obtain integer numbers to order.

6. She thinks she can sell 120 pair of discount Nike tennis shoes to the girls in the school. Justify her reasoning. (Hint: use the survey data)

   There were 300 women who answered the survey as compared to 200 men. This could be an indication that women have more interest in this product than men. The 300 is 50% more. Fifty percent more than 80 men's shoes is 120 women's shoes.

7. How many of each color women's shoe should she stock in the store at Covington High School? Fill in Table 7.

   Multiply the percentages in Table 1 by 120 to obtain the number to order. In this case all of the values were integer. These are recorded in Table 2.

Men			Women		
Color	**Percent**	**# of men**	**Color**	**Percent**	**# of women**
Black	*55%*	*44*	**Black**	8.3%	*10*
White	*11.5%*	*9*	**White**	60%	*72*
Grey	*17.5%*	*14*	**Green**	8.3%	*10*
Blue	*8.5%*	*7*	**Blue**	20%	*24*
Orange	*7.5%*	*6*	**Orange**	3.3%	*4*
Totals	*100%*	80	**Totals**	*100%*	120

Table 7: Tennis shoe color order

The survey also asked the students to report their shoe sizes. Table 8 contains the shoe size data. The data were tabulated and the percentages for each shoe size were calculated.

Men	**Size**	**9 1/2**	**10**	**10 1/2**	**11**	**11 1/2**	**TOTAL**
	Percent	8%	19%	40%	23%	10%	100%
Women	**Size**	**7 1/2**	**8**	**8 1/2**	**9**	**9 1/2**	
	Percent	12%	17%	45%	18%	8%	100%

Table 8: Tennis shoe size data survey

Write four statements that you can make based on Table 3. *Answers will vary. Example: The largest percentage of men wear size 10 ½ and the largest percentage of women wear size 8 ½.*

When placing the order. Bernice new she had to specify the sizes for each color. Take the men's color information in Table 7 and record it in the "Totals" column of Table 9.

8. Next use the information in Table 8 to help Bernice place her men's order.

		Totals	Men's shoe size				
			9 1/2	10	10 1/2	11	11 1/2
			8%	19%	40%	23%	10%
Color	Black	*44*	*3.52*	*8.36*	*17.60*	*10.12*	*4.40*
	White	*9*	*0.72*	*1.71*	*3.60*	*2.07*	*0.90*
	Grey	*14*	*1.12*	*2.66*	*5.60*	*3.22*	*1.40*
	Blue	*7*	*0.56*	*1.33*	*2.80*	*1.61*	*0.70*
	Orange	*6*	*0.48*	*1.14*	*2.40*	*1.38*	*0.60*

Table 9: Men's tennis shoe color order by size

When you multiply the percentages for each shoe size by the color total, you would get the numbers presented above. When you round the values for Black, you obtain the numbers below. These still add up to 44.

	Totals	9 1/2	10	10 1/2	11	11 1/2
Black	*44*	*4*	*8*	*18*	*10*	*4*

However, when you round the values for White, you obtain the numbers below. These do not add up to 9. Because of rounding, the total is now 10. We subtracted 1 from the largest value, size 10 ½. The same rounding problem occurs with Blue. The opposite problem occurs with Orange. It is necessary to add 1 to a specific shoe size. We chose the size with the smallest value:

	Totals	9 1/2	10	10 1/2	11	11 1/2
White (adds 1)	10	*1*	*2*	*4*	*2*	*1*
White (revised down)	9	*1*	*2*	*3*	*2*	*1*
Blue (adds 1)	8	*1*	*1*	*3*	*2*	*1*
Blue (revised down)	7	*1*	*1*	*2*	*2*	*1*
Orange (subtracts 1)	5	*0*	*1*	*2*	*2*	*1*
Orange (revised up)	6	*1*	*1*	*2*	*2*	*1*

Finally, take the women's color information in Table 7 and record it in the "Totals" column of Table 10.

9. Next use the information in Table 8 to help Bernice place her women's order for each color and size.

The results are given in the table below. These values are rounded as shown in the next table. However, rounding changes the totals. These number are then adjusted to keep the original color totals.

		Totals	**Women's shoe size**				
			7 1/2	**8**	**8 1/2**	**9**	**9 1/2**
			12%	17%	45%	18%	8%
Color	**Black**	*10*	*1.2*	*1.7*	*4.5*	*1.8*	*0.8*
	White	*72*	*8.64*	*12.24*	*32.4*	*12.96*	*5.76*
	Grey	*10*	*1.2*	*1.7*	*4.5*	*1.8*	*0.8*
	Blue	*24*	*2.88*	*4.08*	*10.8*	*4.32*	*1.92*
	Orange	*4*	*0.48*	*0.68*	*1.8*	*0.72*	*0.32*

Table 10: Women's tennis shoe color order by size

	Totals	Women's shoe size				
		7 1/2	**8**	**8 1/2**	**9**	**9 1/2**
		12%	17%	45%	18%	8%
Black (rounding, adds 1)	*11*	*1*	*2*	*5*	*2*	*1*
Black (revised down)	*10*	*1*	*2*	*4*	*2*	*1*
White (as is)	*72*	*9*	*12*	*32*	*13*	*6*
Grey (rounding, adds 1)	*11*	*1*	*2*	*5*	*2*	*1*
Grey (revised down)	*10*	*1*	*2*	*4*	*2*	*1*
Blue (as is)	*24*	*3*	*4*	*11*	*4*	*2*
Orange (rounding, adds 1)	*5*	*1*	*1*	*2*	*1*	*0*
Orange (revised down)	*4*	*1*	*1*	*1*	*1*	*0*

Activity 11:
Compound Growth of App Users

Mathematical goals

The student will forecast the growth in the number of users by compounding percentage growth. This example involves multiple steps that combine compounding percentages and some economic forecasts.

The student will:

- Read and use data presented in table format.
- Calculate percentages.
- Compound percentages.
- Interpret graphs.
- Use algebraic expression to determine future growth
- Economic forecasts.

Before the lesson (5-10 minutes)

Put the paper and pencil down and practice some mental mathematics.

Number Talk Possibilities:

Select two or three depending on student abilities.

- $5 \times 5 \times 5 \times 5$
- $2.5 \times 2.5 \times 2.5$
- 6 factors of 2
- any number, n, factors of 3

Growth of app users

Wikipedia calls a mobile app a "software application designed to run on mobile devices such as smartphones and tablet computers." Some apps make it easy to do things such as sharing pictures and videos. Other apps are games. A company that makes Apps earns money by charging to download an app on a smartphone or tablet. They also make money by charging other companies to use the app to advertise.

Three students at Stanford University developed the successful app Snapchat[1]. When it was launched in 2011, it grew very fast. It continues to grow. The graph in Figure 1 shows the number of monthly active users in 2012. The k on the y-axis represents 1,000. Therefore, 500k means 500,000 users. In March 2012, there were less than 100,000. In October 2012, there were over a half million. A month later there were more than a million.

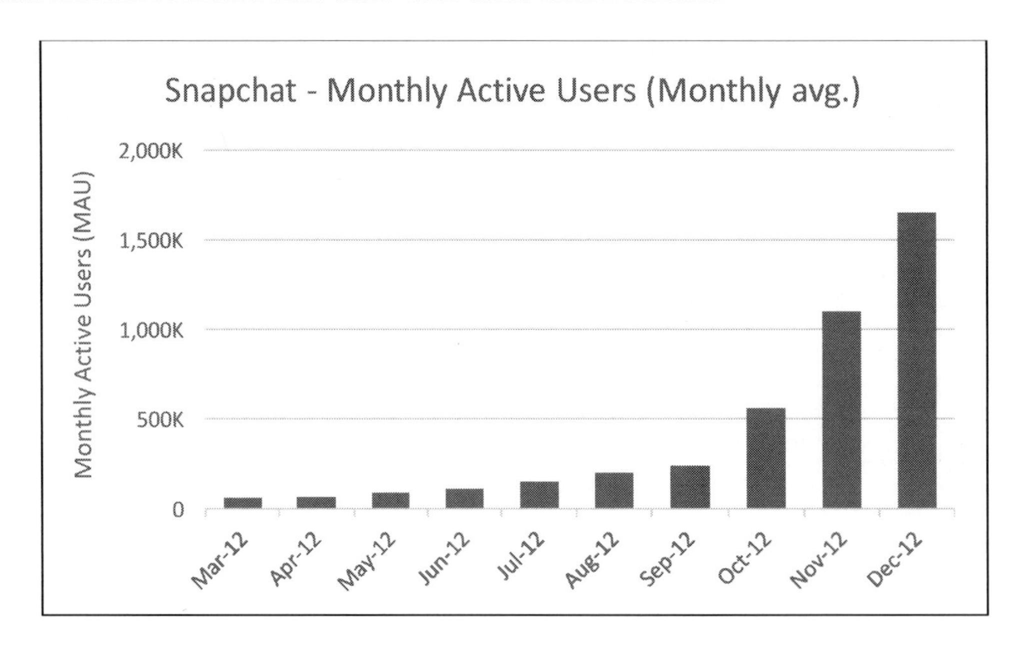

Figure 1: Snapchat Monthly Active Users in 2012

1. Explain how 1000k on the y-axis in Figure 1 represents one million active users.
2. Using Figure 1, about how many monthly active users were there in December 2012?

At first, you could only share pictures on Snapchat. By December of 2012 you could share videos. Today Snapchat estimates that its users view 10 billion videos each day.

[1] https://en.wikipedia.org/wiki/Snapchat

Percentage of game app users

Dante Champion and Brenda Watson just finished the eleventh grade. During the school year, they had taken an app programming course. They decided to develop a game app that makes finding percentages fun and challenging. They worked together. Dante was in charge of designing and testing the app. Brenda was in charge of programming the app. They launched the app during the school year. By the end of June, there were already 1,000 users. As the summer began, they recorded the number of users at the end of each month. Dante calculated the percent growth each month. During July, August, and September, the growth was between 24% and 26% each month.

	Month	Total Users	Actual Percent Growth	Totals Assuming 25% Growth
0	**June**	1000		1000
1	**July**	1258	25.8%	1250
2	**August**	1560	24.0%	1563
3	**September**	1956	25.4%	1953
4	**October**			
5	**November**			
6	**December**			
7	**January**			
8	**February**			

Table 1: Growth in app users

Brenda asked, "How much longer will it take to get to three times the total we had at the end of June?" Dante said they could use an average growth rate of 25% per month to answer the question. He wanted to first see if his method was reasonable. He first used the 25% growth each month to calculate what the totals would have been in the first three months. The numbers were close to the actual values. At the end of three months, with exactly 25% growth each month, the total would have been 1,953. This is close to the actual number, 1,956. So he knew his method of using 25% growth per month to model future growth was reasonable.

However, Brenda was puzzled by the numbers. "If growth was 25% each month, why wasn't the three-month total growth 75%?" she wondered. "There should be only 1,750 users; not 1956." There was almost 100% growth in just three months.

Dante explained to her the idea of "compounding." Each 25% growth applied not only to the original 1,000 but also to the previous month's growth. So, after two months, the total was 156% of the original value. There are two ways to calculate this. First, find the amount of growth in the month and then add it to the original 1,000. The calculation for the first month is:

$$1{,}000 \times 0.25 = 250$$

$$1{,}000 + 250 = 1{,}250$$

This is repeated after the first month to find the second month total.

$$1{,}250 \times 0.25 = 313$$

$$1{,}250 + 313 = 1{,}563$$

Another way to do this calculation can be done in one step by creating a mathematical expression.

Let x = the starting number of users at the end of June.

At the end of July, the number of users is

$$x + 0.25\,x = 1x + 0.25x = 1.25\,x$$

Using the second method, it is easier to see the growth percentage over several months. After two months, the new total is 156% of the original value.

$1.25 \times (1.25\,x) = (1.25 \times 1.25)\,x = 1.56\,x$, if we round off to two decimal places.

After three months, it was 195% of the original total. The increase was 95%.

$1.25 \times 1.25 \times 1.25\,x = 1.95\,x$, again rounded to two decimal places.

3. Help Dante by filling in the last column of Table 1 using the 25% growth model until the total number of users exceeds 4,000.

4. How long should it take to reach 4,000 users if the percentage growth stays the same?

Brenda began to think ahead to when they graduate next June. She wondered how many users there would be at the end of June next year. Dante was getting ready to fill in Table 1 up through month 12. However, Geraldo Wiseacre, a math whiz, said there was an easier way to estimate the number of users at the end of 12 months. He pointed out that repeated multiplication by the same value can be represented by the power function. For example, the number of users at the end of four months would be 1.25 to the fourth power. This is easy to do with a calculator.

$$1.25 \times 1.25 \times 1.25 \times 1.25\,(1{,}000) = 1.25^4\,(1{,}000) = 2{,}441\,(1000) = 2{,}441$$

5. Use the powers of 1.25 to estimate the number of users at the end of 12 months.

Geraldo Wiseacre was also a data whiz. He knew how to use a spreadsheet to do calculations and create graphs. Dante asked Geraldo to create a graph for the next 14 months. This would cover this school year and continue through two summer months. The x-axis would be the months and the y-axis would be the number of users. It should be easy to read. Geraldo created Figure 2.

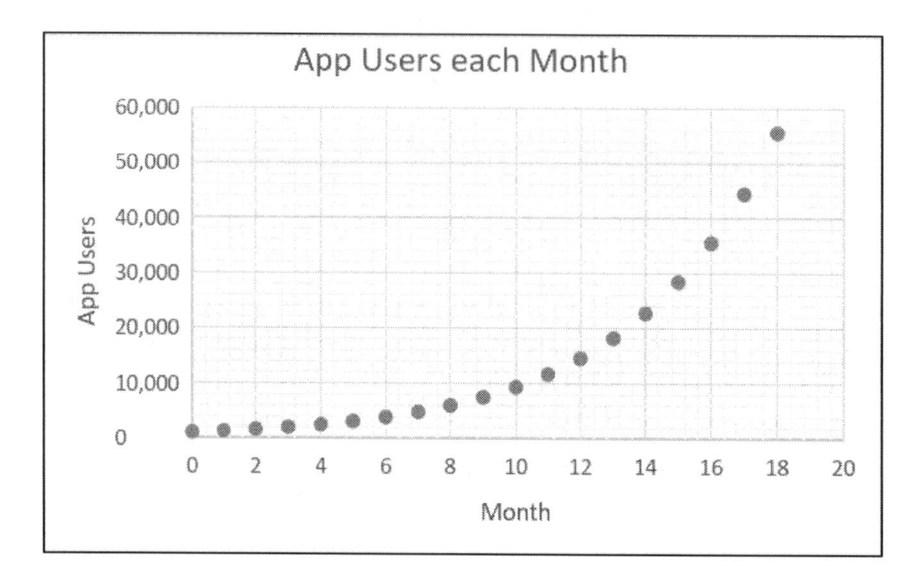

Figure 2: Number of app users each month

After seeing the graph, Brenda noticed the curve got steeper and steeper. It took 5 months to reach about 3,000 total users. However, it took only another 3 months to reach almost 7,000 users at the end of month 8. She could see that it took fewer and fewer months to reach each additional 3,000 user milestone.

6. When will they reach 9,000 users, 12,000 users and 15,000 users? How many months apart are these milestones?

The success of their app was looking good. Brenda and Dante intended to continue to work on it when they went to college together. Both planned to go to the state university that was 40 miles from home. They really did not want to drive back and forth every day. Their families had saved enough money to pay for college tuition. However, they could not afford the cost of living in a dormitory. The cost of room and board for a semester is \$5,000 per student. They also would each need another \$800 per month for living expenses. There were two semesters with four months in each semester. To cover the cost of living in a dormitory for just the fall semester, Dante and Geraldo would each need \$8,200.

Fall semester needs = $5,000 + 4 × $800 = $8,200

Together they would need a combined total of $16,400 for just the fall semester. They would need $32,800 for the two semesters.

They contacted Elsa Prince, an executive at Dalmatian Chocolates. The company had a new product it wanted to advertise. It was a triangle of white chocolate with dark chocolate spots. The product was sold in $4.00 and $7.00 boxes. They asked her how much she would pay to advertise on their app. Elsa offered to start advertising her product once there were at least 3,000 users of the app. She would pay them $0.14 per month for each user at the beginning of each month.

Dante and Brenda were not sure if they would have enough money at the end of the summer to pay for the fall semester. They started calculating how much money they would earn each month. In the first five months they would earn no revenue. They would not yet reach 3,000 users. At the end of November, they expect to have reached more than 3,000 users. They would start earning money from advertising in December. They would earn $427.28 in December.

3,052 × ($0.14) = $427.28

In January, they would earn $534.10

3815 × ($0.14) = $534.10

In total, they would have less than $1,000 at the end of January. They were getting tired of so many tedious calculations. They asked Geraldo to make a graph of the amount of money they expect to earn each month. Figure 3 shows the money they expect to earn each month through the end of the fall semester in college. This was 18 months from the end of last June when they had 1,000 users.

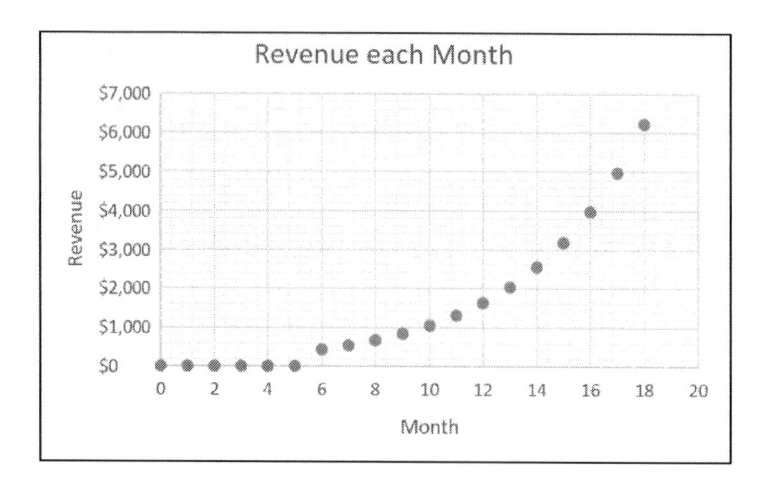

Figure 3: Money earned each month for 18 months

7. Look at Figure 3. In what month would they begin earning more than $800 per month? (This is enough to cover one person's monthly expenses.)

8. Look at Figure 3. In what month would they begin earning more than $1,600 per month? (This is enough to cover both of their monthly expenses).

Brenda and Dante were especially interested in their total earnings. Geraldo produced Figure 4, which shows their total earnings by the end of each month.

9. At the end of August (month 14), will Dante and Brenda have earned enough money to cover each of their first semester's room and board?

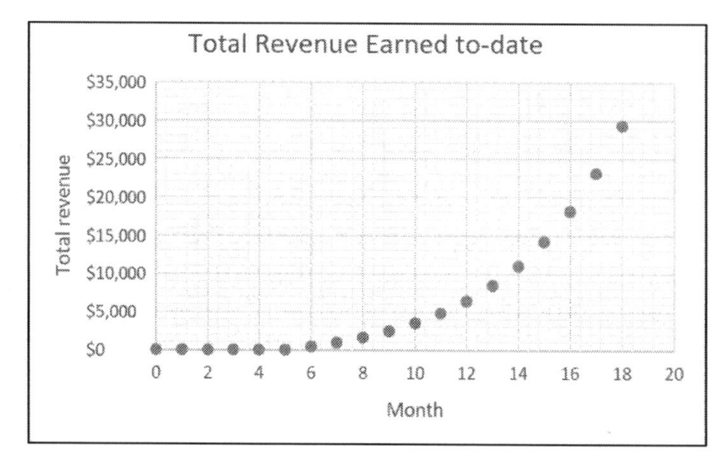

Figure 4: Total revenue earned by the end of each month for 18 months

10. At the end of August (month 14), will Dante and Brenda have earned enough money to cover their fall room and board and their expenses?

11. At the end of December (month 18) will Dante and Brenda have earned enough money to cover all of their expenses for fall and winter semesters?

12. Calculate how much advertising money they will earn in December (month 18).

13. What concerns should Dante and Brenda have about their plans to cover their first year's room and board and other expenses?

Project idea:

Look for a data on the internet about the growth in the number of users for an app or game you are interested in. How fast did the number of users grow in the first few months and first few years? Did it look linear or did it grow faster than linear?

Practice problems

Spread a Rumor

Jimmy, Johnny, Julie and Janie sit together in middle school math class. They tried to imagine mischievous ways to apply a recent lesson on linear and exponential growth. They were thinking of spreading a rumor that school would end after lunch because of a forecast for a major storm beginning in the early afternoon. (FYI There was a major storm coming but it was not going to arrive until late in the evening.)

Jimmy and Johnny wondered how fast the rumor would spread if each of the four members of the group shared the rumor with two different students every 15 minutes.

1. How many new students would hear the rumor in the first 15 minutes? How many new students would hear the rumor in the second 15 minutes?

They wondered how long it would take to reach 100 students not counting themselves. Jimmy created a table and started recording the answers to question 1. He also kept track of the total in the final column.

2. Fill in the values in Table 2.

Time	# of 15 min. segments	New students informed	Total students informed
8:00 am	0	0	0
8:01 am - 8:15 am	1	8	8
8:16 am - 8:30 am	2	8	16
8:31 am - 8:45 am	3		
8:46 am - 9:00 am	4		
9:00 am - 9:15 am	5		

Table 2: 15-minute segments – number of students informed

The process was boring. Johnny said there was an easier way to find the total number of students informed by the end of any interval. He could write a mathematical expression that could be used to figure this out.

3. Let ***n***, represent the number of 15-minute segments. Write an algebraic expression that can be used to calculate directly the total number of students informed for any value of ***n***. Show this expression is correct by calculating the totals for the 5th and 6th time periods.

Johnny also said the mathematical expression could be used to create an algebraic equation. By solving this algebraic equation, he could easily determine how long it would take to reach 100 students

4. Write an equation to determine how long it would take to reach 100 students?

5. Solve the equation and determine how long it would take to reach 100 students. By what time would they reach 100 students?

They were surprised at how long it would take. Jimmy wondered what if they stepped up their game. Each group member talked to 3 students every 15 minutes.

6. Write and solve an equation to determine how long it would take to reach 100 students with this new plan?

The results looked good when three new students were contacted every 15 minutes.

Julie had a better idea. What if they encourage each new student to share the "news" with two more students in just the next time segment? However, Janie thought it would take twice as long to convince a classmate to help spread the rumor. Every 30 minutes each new student would be able to convince just two students. She recorded the numbers for the first hour in Table 3.

Johnny and Jamie laughed. Based on the results, their strategy would reach more students by 9:00 am. However, Julie and Janie were confident their strategy would be faster in getting to 100. The boys' strategy did not get to 100 until 10:15 am, the end of the 9th segment. To speed up the calculations, Julie suggested they input a formula into a calculator.

Time	# of 30 min. segments	New Students Informed	Total Students Informed
8:01 am - 8:30 am	1	$4\times2 = 8$	8
8:31 am – 9:00 am	2	$4\times2\times2 = 16$	24
9:01 am - 9:30 am	3		
9:31 am – 10:00 am	4		
10:00 am – 10:30 am	5		

Table 3: 30-minute segments – number of students informed

7. Let n represent the number of 30-minute time periods. Write an expression that represents the number of students hearing the rumor for the first time during the nth 30 minute segment. Use the calculator to fill in the table.

8. Would Julie and Janie's strategy reach 100 faster than the boys?

9. If you had to choose, which strategy would you use and why?

10. If the middle school population was 824 students, what percentage of the students will have heard the rumor at 10:00 am under the boys' strategy? Under the girls' strategy?

Reduce Mosquito Population

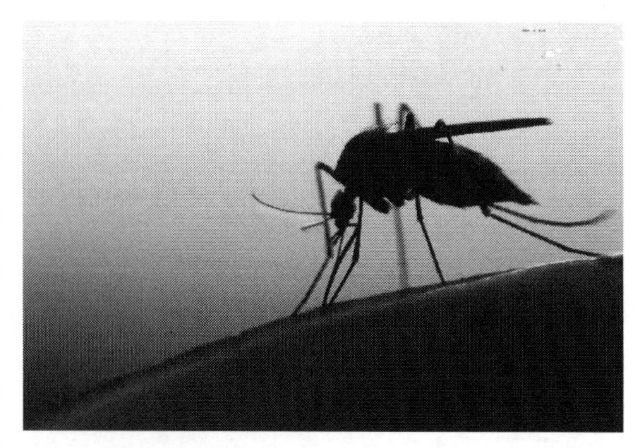

It's the middle of summer and the mosquito population at the local pond has reached an unacceptable level estimated to be 50,000 mosquitos. The DNR authorities have decided to treat the area with a pesticide spray that claims to reduce the population by 35% per application. The DNR approved one application every three days. A population of 2,000 mosquitos is considered an acceptable number for this pond.

11. What is the population of mosquitos before the first application?

12. What is the population of mosquitos after the first application?

13. Assuming the application of pesticide does not allow the mosquitos to reproduce, what is the population of mosquitos after the 4th application?

14. Assuming the application of pesticide does not allow the mosquitos to reproduce, write an expression showing the population remaining after ***n*** applications.

15. Assuming the application of pesticide does not allow the mosquitos to reproduce, how many applications will it take to reduce the population of mosquitos to an acceptable number of 2,000?

Activity 11: Teachers' guide

Thinking through a lesson protocol

Standards:

6. EE.A.1: Write and evaluate numerical expressions involving whole-number exponents.

6.EE.A.2.A: Write expressions that record operations with numbers and with letters standing for numbers.

6.EE.A.2.B: Identify parts of an expression using mathematical terms (sum, term, product, factor, quotient, coefficient); view one or more parts of an expression as a single entity.

6.EE.A.2.C: Evaluate expressions at specific values of their variables. Include expressions that arise from formulas used in real-world problems. Perform arithmetic operations, including those involving whole-number exponents, in the conventional order when there are no parentheses to specify a particular order (Order of Operations).

6.RP.A.3.C: Find a percent of a quantity as a rate per 100 (e.g., 30% of a quantity means 30/100 times the quantity); solve problems involving finding the whole, given a part and the percent.

7.RP.A.3: Use proportional relationships to solve multistep ratio and percent problems.

Mathematical Practices:

MP1: Make sense of problems and persevere in solving them.

MP2: Reason abstractly and quantitatively.

MP3: Construct viable arguments and critique the reasoning of others.

MP4: Model with mathematics.

MP8: Look for and express regularity in repeated reasoning.

Setting up the problem - Launch	
Selecting tasks/goal setting	(10 minutes) Use the video below "One Grain of Rice" to read the story to the class. https://www.youtube.com/watch?v=lxBspOHFMJw
Questions	The teacher may pause periodically throughout the video to ask the students to reflect on what is happening in the story.

Monitoring student work – Explore		
Strategies and misconceptions- Anticipating	**Who - Selecting and sequencing**	**Questions and statements - Monitoring**
(10 Minutes) Read the text and discuss and complete question #1 - 2.		
(10 Minutes) Continue reading about Dante and Brenda and stop after the last full sentence on page 5.		Ask students to calculate, as Dante did, months July, Aug, and Sept on their own. Compare answers to the answers in the table. Discuss any difference that may be caused by rounding and when the rounding is done.
(10 minutes) Before students go on with the reading, ask the students "If it grew 25% a month, why isn't the total growth 75%?" After a brief discussion, have students continue the reading and work questions #3-6 with a partner.		
(20 minutes) Continue reading text and answer questions #7-13 with a partner.		

Monitoring individual student work - Explore		
Strategies and misconceptions - Anticipating	**Who - Selecting and sequencing**	**Questions and statements - Monitoring**
For off-task students or for students that seem to be self-conscious about you listening to them share.		I am just listening or looking to find out how you are working on the problem. This helps me think about what we will do later.
For students that appear to be stuck. Also for when you are having a difficult time understanding their strategies.		Can you tell me a little about your reading? How would you describe the problem in your own words? What facts do you have? Could you try it with simpler numbers?
For students that want to ask you questions, these are ways to uncover their thinking and judge to what extent you want to respond.		Why are you interested in more information about that? Let me say a little about that part. Tell me what you've thought about so far. What do you know?

Managing the discussion – Summarize	
Parts of discussion - Connecting	**Questions and statements - Connecting**
Launching the discussion: Select the problems in questions #6-13 that students are struggling with or you wish to share out.	Will team 1 start us off by sharing one way of working on this problem? Please raise your hand when you are ready to share your solution. What did you do first when you were working on this problem? Let's start by clearing up a few things about the problem. Let's list some key parts in this problem. What was unclear in the problem?
Eliciting and uncovering student strategies	Joe would you be willing to start us off? What have you found so far? Can you repeat that? Can you explain how you got that answer? How do you know? Walk us through your steps. Where did you begin? Can you show us?
Focusing on mathematical ideas	Can you explain why this is true? Does this method always work? How is Bob's method similar to Kelly's method? What do all the solutions have in common? What would happen if I changed the numbers to _____?
Encouraging interactions	Do you agree or disagree with Kahlil's idea? What do others think? Would someone be willing to repeat what Tom just said? Would anyone be willing to add on to what Sue just said?
Concluding the discussion	Can anyone tell me some of the big ideas that we learned today? How would you explain what we learned today to a 5th grader? Some of the key points from our discussion today are . . . Tomorrow we will continue our exploration of ______ beginning with the idea from today that __________.
Post lesson notes	You may wish to assign the practice problems that you feel would benefit the students.

Solutions to text questions

1. Explain how 1000k on the y-axis in Figure 1 represents one million active users.

   1k represents 1000 active users so 1000k represents one thousand times as many users or 1,000,000 active users.

2. Using Figure 1, about how many monthly active users were there in December 2012?

   There are approximately 1700k or 1,700,000 active users in December 2012.

3. Help Dante by filling in the last column of Table 1 using the 25% growth model until the total number of users exceeds 4,000.

Month		Total Users	Actual Percent Growth	Totals Assuming 25% Growth
0	**June**	1000	0	1000
1	**July**	1258	25.8%	1250
2	**August**	1560	24.0%	1563
3	**September**	1956	25.4%	1953
4	**October**			*2,441*
5	**November**			*3,052*
6	**December**			*3,815*
7	**January**			*4,768*
8	**February**			*5,960*
9	***March***			*7,451*
10	***April***			*9,313*
11	***May***			*11,642*
12	***June***			*14,552*
13	***July***			*18,190*
14	***August***			*22,737*
15	***September***			*28,422*
16	***October***			*35,527*
17	***November***			*44,409*
18	***December***			*55,511*

Table 1: Growth in app users

4. How long should it take to reach 4,000 users if the percentage growth stays the same?

   Sometime in the 7th month or January. The end of January indicates 4,768 active users.

5. Use the powers of 1.25 to estimate the number of users at the end of 12 months.

   (1.25^12) × 1000 = 14,551.915 OR 14,552 active users

6. When will they reach 9,000 users, 12,000 users and 15,000 users? How many months apart are these milestones? (Student need to extend the data in the table to answer these questions).

   The 9,000 users will occur sometime in April (the tenth month), the 12,000 users will occur in June, the 15,000 users will occur in July.

7. Look at Figure 3. In what month would they begin earning more than $800 per month? (This is enough to cover one person's monthly expenses.)

   March (the ninth month). Please note that student answers may vary by a month as data is extrapolated from the graph.

8. Look at Figure 3. In what month would they begin earning more than $1,600 per month? (This is enough to cover both of their monthly expenses).

   June (the twelfth month) Please note that student answers may vary by a month as data is extrapolated from the graph.

9. At the end of August (month 14), will Dante and Brenda have earned enough money to cover each of their first semester's room and board?

   Refer to Figure 4.

10. At the end of August (month 14), will Dante and Brenda have earned enough money to cover their fall room and board and their expenses?

    No, it appears on the graph that their revenue is between $10,000 and $12,000. They need $16,400 to cover their expenses so they fall short.

11. At the end of December (month 18) will Dante and Brenda have earned enough money to cover all of their expenses for fall and winter semesters?

    No, it appears on the graph that their revenue is under $30,000. They need $32,800 to meet their fall and winter semesters so they fall short.

12. Calculate how much advertising money they will earn in December (month 18).

    Number of users from November x Revenue per active user = Revenue in December
    44,409 × $0.14 = $6,217.26

13. What concerns should Dante and Brenda have about their plans to cover their first year's room and board and other expenses?

    The plan falls short for the first year's expenses, however, the amount of revenue each month is growing rapidly, allowing future payments to be covered.

Month		Total Users	Actual Percent Growth	Totals Assuming 25% Growth
0	**June**	1000	0	1000
1	**July**	1258	25.8%	1250
2	**August**	1560	24.0%	1563
3	**September**	1956	25.4%	1953
4	***October***	*2,441*	*0*	*0*
5	***November***	*3,052*	*0*	*0*
6	***December***	*3,815*	*$427*	*$427*
7	***January***	*4,768*	*$534*	*$961*
8	***February***	*5,960*	*$668*	*$1,629*
9	***March***	*7,451*	*$834*	*$2,463*
10	***April***	*9,313*	*$1,043*	*$3,506*
11	***May***	*11,642*	*$1,304*	*$4,810*
12	***June***	*14,552*	*$1,630*	*$6,440*
13	***July***	*18,190*	*$2,037*	*$8,477*
14	***August***	*22,737*	*$2,547*	*$11,024*
15	***September***	*28,422*	*$3,183*	*$14,207*
16	***October***	*35,527*	*$3,979*	*$18,186*
17	***November***	*44,409*	*$4,974*	*$23,160*
18	***December***	*55,511*	*$6,217*	*$29,377*

Solutions to practice problems

1. How many new students would hear the rumor in the first 15 minutes? How many new students would hear the rumor in the second 15 minutes?

   Eight (8) new students would hear the rumor in the first 15 minutes, an additional 8 students would hear the rumor in the second 15 minutes.

2. Fill in the values in Table 2.

Time	# of 15 min. segments	New students informed	Total students informed
8:00 am	0	0	0
8:01 am - 8:15 am	1	8	8
8:16 am - 8:30 am	2	8	16
8:31 am - 8:45 am	3	*8*	*24*
8:46 am - 9:00 am	4	*8*	*32*
9:00 am - 9:15 am	5	*8*	*40*

Table 2: 15-minute segments – number of students informed

3. Let ***n***, represent the number of 15-minute segments. Write an algebraic expression that can be used to calculate directly the total number of students informed for any value of ***n***. Show this expression is correct by calculating the totals for the 5th and 6th time periods.

   Total number of students informed is equal to the number of 15-minute segments times 8. 8n. For 5th time period, $8 \times 5 = 40$ and for 6th time period, $8 \times 6 = 48$.

4. Write an equation to determine how long it would take to reach 100 students?

   $8n = 100$

5. Solve the equation and determine the how long it would take to reach 100 students. By what time would they reach 100 students?

   $8n = 100$; $n = 100 / 8$ or 12.5. It would take 12.5 fifteen-minute segments for 100 students to be informed. 12 segments would not be enough so you would need 13 segments. 13 fifteen-minute segments is 195 minutes or 3 hours and 15 minutes after 8:00 am or by 11:15 am.

They were surprised at how long it would take. Jimmy wondered what if we stepped up our game and each talked to 3 students every 15 minutes.

6. Write and solve an equation to determine how long it would take to reach 100 students with this new plan?

   4 team members each sharing with 3 students would mean 12 new students informed every 15 minutes. 12n = 100; n = 8.3333 It would take 9 fifteen-segments to reach 100 students. It would take until 10:15 am.

Time	# of 30 min. segments	New Students Informed	Total Students Informed
8:01 am - 8:30 am	1	$4 \times 2 = 8$	8
8:31 am – 9:00 am	2	$4 \times 2 \times 2 = 16$	24
9:01 am - 9:30 am	3	*$4 \times 2 \times 2 \times 2 = 32$*	*56*
9:31 am – 10:00 am	4	*$4 \times 2 \times 2 \times 2 \times 2 = 64$*	*120*
10:00 am – 10:30 am	5	*$4 \times 2 \times 2 \times 2 \times 2 \times 2 = 128$*	*248*

Table 3: 30-minute segments – number of students informed

7. Let ***n*** represent the number of 30-minute time periods. Write an expression that represents the number of students hearing the rumor for the first time during the nth 30 minute time period. Use the calculator to fill in Table 3.

   4(2n). See Table 3.

8. Would Julie and Janie's strategy reach 100 students faster than the boys' strategy?

   Yes, 120 students would know by 10:00 am.

9. If you had to choose which strategy would you use and why?

   Answers will vary. The girls' strategy reaches more students faster but relies on new students sharing with other student who have not yet heard the rumor. If they were the only ones sharing the rumor, they would know who had and had not been told.

10. If the middle school population was 824 students, what percentage of the students will have heard the rumor at 10:30 am under the boys' strategy? Under the girls' strategy?

    Boys' strategy would include 10 fifteen-minute segments; 12 × 10 = 120; 120 / 824 = 0.145631 or 14.6%. The girls' strategy would include 5 thirty-minute segments; looking at the table 248 total students would be informed 248/824 = 0.30097 or 30.1%

11. What is the population of mosquitos before the first application?

    50,000 mosquitos

12. What is the population of mosquitos after the first application?

    $50{,}000 \times 0.65 = 32{,}500$ *mosquitos*

13. Assuming the application of pesticide does not allow the mosquitos to reproduce, what is the population of mosquitos after the 4th application?

    $50{,}000 \times 0.65 \times 0.65 \times 0.65 \times 0.65 = 8{,}925$ *mosquitos*

14. Assuming the application of pesticide does not allow the mosquitos to reproduce, write an expression showing the population remaining after ***n*** applications.

    $50{,}000(0.65n)$

15. Assuming the application of pesticide does not allow the mosquitos to reproduce, how many applications will it take to reduce the population of mosquitos to an acceptable number of 2,000?

    See table below. After 7 applications, there would still be 2,451 mosquitos, so you would need 8 applications to reduce the population of mosquitos to under 2,000.

Amount after Application	***50000***
1	*32500*
2	*21125*
3	*13731.25*
4	*8925.313*
5	*5801.453*
6	*3770.945*
7	*2451.114*
8	*1593.224*
9	*1035.596*

Activity 12: Reducing the Number of Homeless Veterans

Mathematical goals

The student will forecast the decline in the number of homeless veterans by compounding percentage decline. The example involves comparing compounding percentages with fixed rate of decrease.

The student will:

- Read and use data presented in table format
- Calculate percentages
- Compound percentages
- Interpret graphs
- Choose between programs

Before the lesson (5-10 minutes)

Put the paper and pencil down and practice some mental mathematics.

Number Talk Possibilities:

Select two or three depending on student abilities.

- If 25% are walking daily, what percent are not walking daily?
- If 14% are bringing their lunch, what percentage are not bringing their lunch?
- Write the expression (0.43×0.43×0.43×0.43) using an exponent.
- Write the expression (2.5×2.5×2.5×2.5×2.5) using an exponent.

Percentage decline

Homelessness is a national problem. HUD (Department of Housing and Urban Development) must present an annual report to Congress on the current situation. In January of each year, HUD carries out a census, called a Point-in-Time, to find the number of homeless people on a single night. In 2015 HUD reported there were 564,708 homeless people. Of these, 69% had found a sheltered place to sleep. The other 31% had only an unsheltered place to sleep. Unsheltered Homeless People stay in places not meant for human habitation. For example, they might sleep on the street or in vehicles, parks, or abandoned buildings. Sheltered Homeless People stay in emergency shelters, transitional housing programs, or other safe havens. Homelessness decreased by 2% between 2014 and 2015 and by 11% since 2007.

Homeless veterans have drawn special attention. The Veterans Administration has worked effectively with HUD to reduce the number of homeless veterans. In 2015 there were 47,725 homeless veterans. This was down 35% since 2009. In the January 2016 census, HUD found there was another 17% decline since 2015. The 2016 total was 39,471. Approximately one-third of these homeless veterans, 13,067, had to sleep unsheltered at night[1].

Which program is the better way to reduce homelessness?

Not all parts of the United States had a 17% decrease in the number of homeless veterans during 2015. In Hawaii, Oregon and Utah, the decreases were less than 10%. Even worse, in Colorado, Louisiana, Oklahoma and South Carolina, the number of homeless veterans increased.

Sarita Olive is the official in charge of addressing the homeless problem in a southern state. Among the 7400 homeless there were 600 homeless veterans. Because of their service to the country, there are special programs to help veterans. Sarita wants to reduce homelessness in her state. She must choose one of two programs that have been effective. One program is a complete approach to helping veterans get their lives in order. It had been shown to reduce homelessness by 17% each year. The other program is focused on building or finding an additional 75 permanent residences each year.[2]

Isaac Green was in charge of data analysis and planning for the homeless project. He compared the two programs. He estimated that in the first year with the Complete Program, 102 veterans out of 600 would find places to live:

---

[1] The above data come from The 2015 Annual Homeless Assessment Report (AHAR) to Congress. https://www.hudexchange.info/resources/documents/2015-AHAR-Part-1.pdf

[2] In this example the number of veterans still homeless at the end of each year were those homeless who did not find residences that year. The actual situation is more complicated. In fact, more homeless people find residences each year. However, each year there are also new individuals who become homeless. The programs described above are measured by the difference in the total number of homeless from year to year.

Number who found places to live is 17% of 600 or 102

At the end of the year, there would still be 498 homeless veterans

Number still homeless is $600 - 102 = 498$

In the second year with the Complete Program, 17% of the remaining 498 homeless would find permanent places to live that year. A total of 85 veterans would find places in the second year. This would leave 413 veterans still homeless. Over the two years, 187 veterans, under this program, would have found permanent homes.

Number who found places to live in the second year is $0.17 \times 498 = 85$

Total number over two years who found places to live is $102 + 85 = 187$

Number still homeless after two years is $498 - 85 = 413$

There is another way to calculate the number still homeless with the Complete Program. If 17% are no longer homeless, that leaves 83% still homeless. The number of homeless after one year is

Homeless after year 1 is 0.83×600

The number of homeless after year 2 is

Homeless after year 2 is $0.83 \times (0.83 \times 600) = 0.83^2 \times 600 \approx 0.689 \times 600 = 413$

Isaac told Sarita that she could write an algebraic expression to represent this calculation. However, this expression is different than the linear expressions they learned in beginning algebra. In this case, the variable is an exponent.

Let: n = the number of years

Then: the number of homeless at the end of year $n = (0.83)^n(600)$

Isaac explained that if Sarita wanted to know the number of homeless after year 3, she could substitute the number 3 in place of the variable n.

The number of homeless after year 3 is $(0.83)^3(600) \approx 0.572(600) = 343$

Sarita wanted to know the number of homeless after year 5, so she replaced the variable n with the number 5.

The number of homeless after year 5 is $(0.83)^5(600) \approx 0.394(600) = 236$

1. Determine the number of homeless under the Complete Program at the end of years 4 and 6 and record the results in Table 1.

	Complete Program - 17% decrease per year		
Year	**Remaining Homeless**	**Decrease in Homelessness for the Year**	**Total Decrease**
Current	600		
1	498	102	102
2	413	85	187
3	343		
4			
5	236		
6			

Table 1: Predicted number of homeless veterans under the Complete Program

The number of veterans placed in homes each year is equal to the change in the number of homeless. At the end of year 1, there would be 498 homeless. At the end of year 2, there would be 413 homeless. This is 85 less than the year before. Thus, 85 fewer veterans would be homeless in year 2 under this program.

$498 - 413 = 85$

2. How many fewer veterans would be homeless during year 3?

3. How many fewer veterans would be homeless during years 3, 4, 5 and 6? Record the answers in Table 1.

One final number that is tracked is the fewer number of veterans who are homeless. By the end of year 2 this would be 187. This is just the difference between the starting number of 600, and 413, the number still homeless after year 2.

$600 - 413 = 187$

4. What is the total number of veterans placed in residences by the end of year 3, 4, 5 and 6? Record the answers in Table 1.

Under the Residence Program, there would be 75 fewer homeless each year. After 1 year, there would be 525 homeless veterans.

$600 - 75 = 525$

After year 2, there would still be 450 homeless veterans.

$525 - 75 = 450$

However, the number that would remain homeless after year 2 can be calculated directly from the initial number. After two years, 75×2 or 150 would have found homes.

$600 - (75 \times 2) = 600 - 150 = 450$

Let n = the number of years The Residence Program has been used

5. Write an algebraic expression to represent the number of homeless veterans at the end of year n under the Residence Program?

6. Under the Residence Program, how many veterans would still be homeless after year 3?

7. How many fewer veterans would be homeless during year 3?

8. Calculate the number homeless at the end of year 4, 5 and 6 and record in Table 2?

Residence Program - 75 Residences Per Year			
Year	**Remain Homeless**	**Placed in residences during the year**	**Total Placed**
Current	600		
1	525	75	75
2	450	75	150
3	375	75	
4			
5			
6			450

Table 2: The Residence Program and the number of homeless veterans

In Table 2, there is a column heading, "Placed in residences during the year."

9. Why is this column easy to fill in with the Residence Program?

The Total Placed at the end of any year can be calculated directly. For example, at the end of year 6, the number placed is six times 75.

$6 \times 75 = 450$

10. Write an algebraic expression to represent the number of veterans that have been helped by the end of year n under the Residence Program?

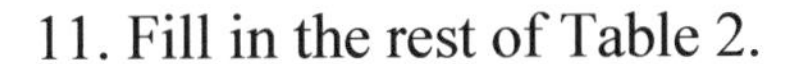

11. Fill in the rest of Table 2.

Sarita asked Isaac to create graphs that would show the two programs side-by-side. Figure 1 shows the predicted number of homeless each year with each program.

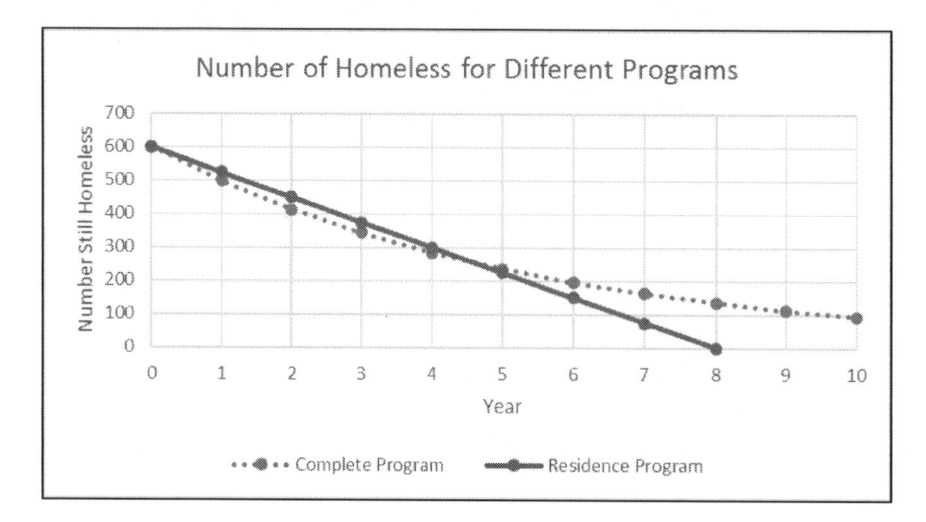

Figure 1: The number of homeless each year for the Complete and Residence Programs

Sarita's state was provided with enough funds to run one of the programs for just three years.

12. Which program should she use if the program is running for just three years?

Sarita was trying to convince senior management to provide funding for six years.

13. If she gets six years of funding which program should she use?

14. At approximately what point on the x-axis do the graphs in Figure 2 intersect? What does this represent?

15. At approximately what point on the y-axis do the graphs in Figure 2 intersect? What does this represent?

Isaac suggested that Sarita had not carefully thought through all of the choices. What if they ran The Complete Program for a few years and then switched to The Residence Program? (Programs must be run for whole years. Sarita cannot change programs in the middle of a year.)

Sarita thought that might be a good idea. She asked Isaac to make another graph that showed the number of fewer homeless veterans at the end of each year. Figure 2 compares the two programs.

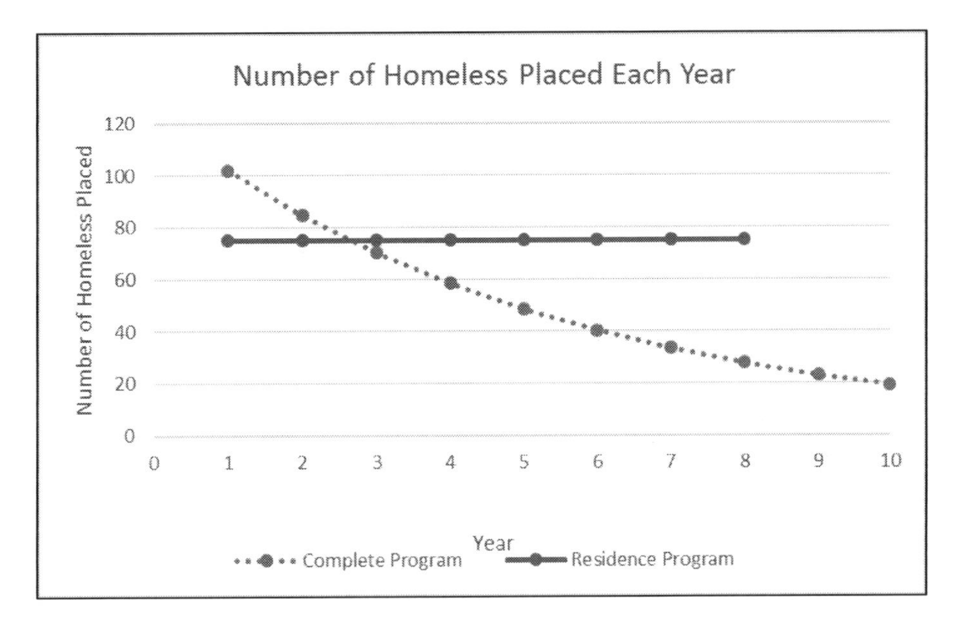

Figure 2: The number of homeless placed each year for the Complete and Residence Programs

16. In Figure 2 the line for Residence Program is parallel to the x-axis. Why is that?

17. Based on Figure 2, which program should they start with? How many years should they use this program before switching to the other program?

18. Refer back to Tables 1 and 2 to explain how the data in the two tables supports the recommendation to switch?

19. Explain why Sarita might decide to use just one policy and stick with it.

Project idea:

Look at data on homelessness in your city and state. Is the number growing or declining over the past five years? What percent of the homeless population are veterans? What are some of the programs that are used to reduce the homeless population in your area?

Practice problems

A survey of Hardy area teens determined that 500 are looking for employment but have been unable to find jobs. The state employment agency is reviewing two programs that can reduce teenage unemployment. However, they have enough money to support only one program for 15 months. Hardy's local schools offer a three-month training program that will result in jobs for 45 teens each quarter. Local churches[3] in Hardy have submitted a proposal for reducing local unemployment. Their program uses their social network to reach out to area companies and stores. They would commit to placing 12% of the unemployed teens in jobs each quarter or every three months.

1. Under the Hardy school's program, how many teens will still be unemployed at the end of the first quarter? At the end of the second quarter?

2. Write an expression for the number of teens that would remain unemployed at the end of ***n*** quarters.

[3] Religious organizations can receive money to provide social services.

3. Complete Table 3 for the school program.

Local School Program - 45 Employed Per Quarter			
Year	**Number who found jobs for the quarter**	**Total placed in jobs**	**Remaining unemployed teens**
Current			500
1	45	45	455
2			
3			
4			
5			
6			

Table 3: Hardy school's program - 45 each quarter

4. Under the Hardy area church program, how many teens will be placed in jobs at the end of the first quarter? At the end of the second quarter?

5. Write an expression for the total number of teens that would be placed in jobs at the end of n quarters.

6. Complete Table 4 for the Hardy area church run program.

Quarter	Remaining unemployed teens	Number who found jobs for the quarter	Total placed in jobs
Current	500		
1	440	60	60
2			
3			
4			
5			

Table 4: Hardy area church run program - 12% each quarter

7. Which of these programs will have a larger decrease in the total number of unemployed teens at the end of the next five quarters? Explain your reasoning.

8. Create a graph for the school run program over a five quarter period.

9. Create a graph for the church run program over a five quarter period.

Activity 12: Teachers' guide

Thinking through a lesson protocol

Standards:

6. EE.A.1: Write and evaluate numerical expressions involving whole-number exponents.

6.EE.A.2.A: Write expressions that record operations with numbers and with letters standing for numbers.

6.EE.A.2.B: Identify parts of an expression using mathematical terms (sum, term, product, factor, quotient, coefficient); view one or more parts of an expression as a single entity.

6.EE.A.2.C: Evaluate expressions at specific values of their variables. Include expressions that arise from formulas used in real-world problems. Perform arithmetic operations, including those involving whole-number exponents, in the conventional order when there are no parentheses to specify a particular order (Order of Operations).

6.RP.A.3.C: Find a percent of a quantity as a rate per 100 (e.g., 30% of a quantity means 30/100 times the quantity); solve problems involving finding the whole, given a part and the percent.

7.RP.A.3: Use proportional relationships to solve multistep ratio and percent problems.

Mathematical Practices:

MP1: Make sense of problems and persevere in solving them.

MP2: Reason abstractly and quantitatively.

MP3: Construct viable arguments and critique the reasoning of others.

MP4: Model with mathematics.

Setting up the problem - Launch	
Selecting tasks/goal setting	(10 minutes) Have students visit the homeless website and record at least two pieces of data with percentages.
Questions	https://www.hudexchange.info/resources/documents/2015-AHAR-Part-1.pdf

Monitoring student work - Explore		
Strategies and misconceptions-Anticipating	**Who - Selecting and sequencing**	**Questions and statements - Monitoring**
(20 Minutes) Read the first two paragraphs in the text. Continue reading the text "Which Program is the Better Way to Reduce Homelessness?" (The teacher needs to record the expressions and mathematics on the board as they go through this text).		Teacher may want to show calculations for the number of homeless after year two leading to using a variable for the exponent.
(10 Minutes) Answer questions #1 and #2.		Students compare their answers with their partners.
(10 minutes) Answer questions #3, #4, #5.		
(20 minutes) Have students read the text on the Residence Program. Answer questions #6 -#12 which completes the table on the Residence Program.		If the teacher has the table on the board, have students come up and complete it. Clarify any questions. Look at the graph "Number of Homeless for Different Programs". Explain the key features of the graph. What does the blue curve represent? What does the y axis represent? What are the units of each of the axes? etc.
(20 minutes) Continue reading the text after question #12. Answers questions #13 thru #16.		Summarize student findings in regards to the comparison of the two plans.
Continue reading the text after question #16. Answer questions #17 thru #20.		Use the terms constant rate of change variable (non-constant) rate of change.

Monitoring individual student work - Explore		
Strategies and misconceptions - Anticipating	**Who - Selecting and sequencing**	**Questions and statements - Monitoring**
For off-task students or for students that seem to be self-conscious about you listening to them share.		I am just listening or looking to find out how you are working on the problem. This helps me think about what we will do later.
For students that appear to be stuck. Also for when you are having a difficult time understanding their strategies.		Can you tell me a little about your reading? How would you describe the problem in your own words? What facts do you have? Could you try it with simpler numbers?
For students that want to ask you questions, these are ways to uncover their thinking and judge to what extent you want to respond.		Why are you interested in more information about that? Let me say a little about that part. Tell me what you've thought about so far. What do you know?

Managing the discussion - Summarize	
Parts of discussion - Connecting	**Questions and statements - Connecting**
Launching the discussion: Select the problems in questions #13-20 that students are struggling with or you wish to share out.	Will team 1 start us off by sharing one way of working on this problem? Please raise your hand when you are ready to share your solution. What did you do first when you were working on this problem? Let's start by clearing up a few things about the problem. Let's list some key parts in this problem. What was unclear in the problem?
Eliciting and uncovering student strategies	Joe would you be willing to start us off? What have you found so far? Can you repeat that? Can you explain how you got that answer? How do you know? Walk us through your steps. Where did you begin? Can you show us?
Focusing on mathematical ideas	Can you explain why this is true? Does this method always work? How is Bob's method similar to Kelly's method? What do all the solutions have in common? What would happen if I changed the numbers to _____?
Encouraging interactions	Do you agree or disagree with Kahlil's idea? What do others think? Would someone be willing to repeat what Tom just said? Would anyone be willing to add on to what Sue just said?
Concluding the discussion	Can anyone tell me some of the big ideas that we learned today? How would you explain what we learned today to a 5th grader? Some of the key points from our discussion today are . . . Tomorrow we will continue our exploration of ______ beginning with the idea from today that __________.
Post lesson notes	You may wish to assign the practice problems that you feel would benefit the students.

Solutions to text questions

1. Determine the number of homeless under the Complete Program at the end of years 4 and 6 and record the results in Table 1.

	Complete Program - 17% decrease per year		
Year	**Remaining Homeless**	**Decrease in Homelessness for the Year**	**Total Decrease**
Current	600		
1	498	102	102
2	413	85	187
3	343	*413 – 343=70*	*187 + 70=257*
4	$0.83 \times 600=284$	*343 – 284=59*	*257 +59 = 316*
5	236	*284 – 236=48*	*316 + 48=364*
6	$0.83^6 \times 600=196$	*236 – 196=40*	*364 + 40=404*

Table 1: The predicted number of homeless veterans under the Complete Program

2. How many fewer veterans would be homeless during year 3?

   413 – 343=70

3. How many fewer veterans would be homeless during years 3, 4, 5 and 6? Record the answers in Table 1.

   See column three in table above.

4. What is the total number of veterans placed in residences by the end of year 3, 4, 5 and 6? Record the answers in Table 1.

   See column four in table above.

5. Write an algebraic expression to represent the number of homeless veterans at the end of year n under the Residence Program?

   $600 - 75n$

6. Under the Residence Program, how many veterans would still be homeless after year 3?

   $600 - 75(3) = 375$

7. How many fewer veterans would be homeless during year 3 in the Residence Program?

   75 veterans would have found homes because every year there are 75 fewer veterans homeless under the Residence Program.

8. Calculate the number homeless at the end of year 4, 5 and 6 and record in Table 2?

Residence Program - 75 Residences Per Year			
Year	**Remain Homeless**	**Placed in residences during the year**	**Total Placed**
Current	600		
1	525	75	75
2	450	75	150
3	375	75	*225*
4	*300*	*75*	*300*
5	*225*	*75*	*375*
6	*150*	*75*	450

Table 2: The Residence Program and the number of homeless veterans

In Table 2, there is a column heading, "Placed in residences during the year."

9. Why is this column easy to fill in with the Residence Program?

Because the Residence Program places 75 homeless veterans each year of the program. It remains constant.

10. Write an algebraic expression to represent the number of veterans that have been helped by the end of year n under the Residence Program?

75n

11. Fill in the rest of Table 2.

See Table 2 above.

12. Which program should she use if the program is running for just three years?

There would be fewer homeless under the Complete Program (blue on graph) at the end of three years. She should use the Complete Program.

13. If she gets six years of funding which program should she use?

The Residence Program has fewer homeless at the end of six years. She should use the Residence Program.

14. At approximately what point on the x-axis do the graphs in Figure 2 intersect? What does this represent?

The graphs intersect at approximately 5 years. This means that at the end of 5 years the number still homeless is approximately the same under either program.

15. At approximately what point on the y-axis do the graphs in Figure 2 intersect? What does this represent?

The graphs intersect at approximately 225. This represents approximately the same number still homeless under either program.

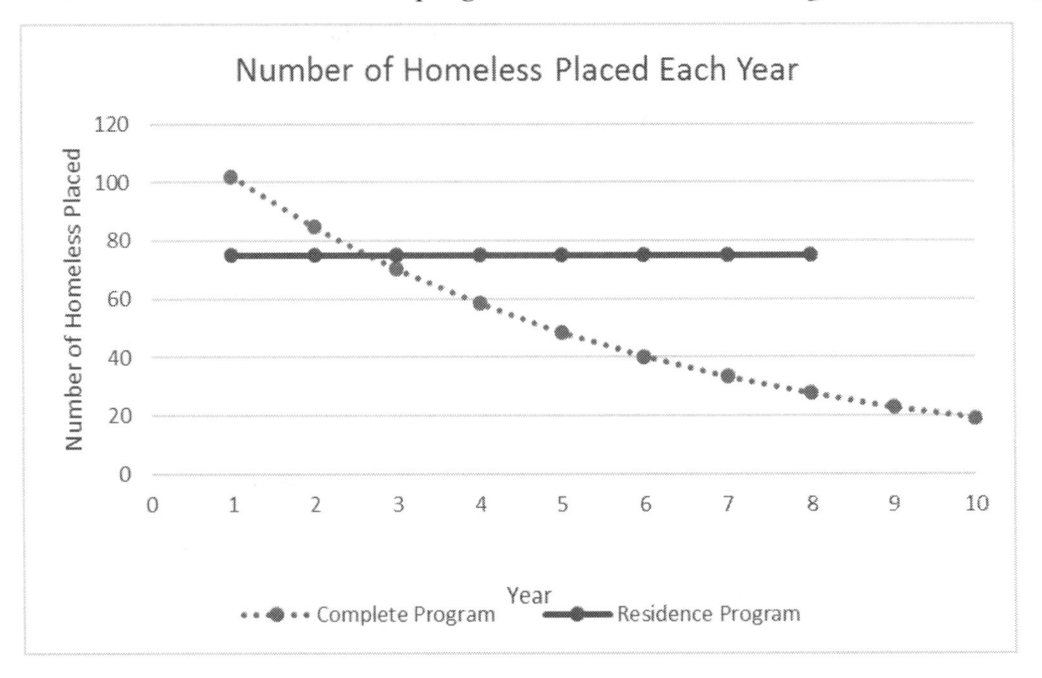

Figure 2: The number of homeless placed each year for the Complete and Residence Programs

16. In Figure 2 the line for Residence Program is parallel to the x-axis. Why is that?

    The line is parallel to the x-axis because the number finding homes each year, 75, remains constant.

17. Based on Figure 2, which program should they start with? How many years should they use this program before switching to the other program?

    Based on Figure 2, they should start with the Complete Program where the number of homeless placed is over 100 instead of 75 for the Residence Program. At the end of year 3 they should switch to the Residence Program which is 75 less homeless and remaining constant where the Complete Program is approximately 70 homeless placed and continuing to decline.

18. Refer back to Tables 1 and 2 to explain how the data in the two tables supports the recommendation to switch?

    According to the tables, 102 are placed in Year 1 in the Complete Program and only 75 in the Residence Program. An extra 77 are placed in Complete Program. In Year 2, 85 are placed in the Complete Program and only 75 in the Residence Program.

An extra 10 are placed in Complete Program. In Year 3, 70 are placed in the Complete Program and 75 are placed in the Residence Program. Five less are placed in the Complete Program indicating a change is desired from the Complete Program to the Residence Program.

Explain why Sarita might decide to use just one policy and stick with it.

Even though this would not be the best situation for years one and two, Sarita might decide to simply adopt the Residence Program from the start, so that she would not have to change programs after year 3.

Solutions to practice problems

1. Under the Hardy school's program, how many teens will be placed in jobs at the end of the first quarter?

   45

   At the end of the second quarter?

   45

2. Write an expression for the total number of teens that would be placed in jobs at the end of n quarters.

   $45n$

3. Complete the Table 3 for the school program.

Local School Program - 45 Employed Per Quarter			
Year	**Number who found jobs for the quarter**	**Total placed in jobs**	**Remaining unemployed teens**
Current			500
1	45	45	455
2	*410*	*45*	*90*
3	*365*	*45*	*135*
4	*320*	*45*	*180*
5	*275*	*45*	*225*

Table 3: Hardy school's program - 45 each quarter

4. Under the Hardy area church run program, how many teens will still be unemployed at the end of the first quarter?

   If 12% find jobs, 88% remain without jobs. $0.88 \times 500 = 440$ *unemployed teens*

   At the end of the second quarter?

   $0.88 \times 0.88 \times 500 = 387.2$ *or 387 teens still unemployed OR* $500 - (0.12 \times 500) = 440$ *and* $500 - (0.12 \times 0.12 \times 500) = 387.2$

5. Write an expression for the number of teens that would remain unemployed at the end of n quarters.

   $500(0.88n)$ *OR* $500 - (0.12n)(500)$

6. Complete Table 4 for the local church program.

   Depending on the rounding process used, answers may vary slightly. This is rounded to the nearest whole number.

Quarter	Remaining unemployed teens	Number who found jobs for the quarter	Total placed in jobs
Current	500		
1	440	60	60
2	*387*	*53*	*113*
3	*341*	*46*	*159*
4	*300*	*41*	*200*
5	*264*	*36*	*236*

Table 4: Hardy area church program - 12% each quarter

7. Which of these programs will have a larger decrease in the total number of unemployed teens at the end of the next five quarters? Explain your reasoning.

   The church program will have a total decrease of 236 teens at the end of five quarters. The local school program will have a total decrease of 225 teens at the end of the five quarters. The local church program placed 11 more teens by the end of the fifth quarter.

8. Create a graph for the Hardy school's program over a five quarter period.

   A graph could be created by comparing number of quarters vs number of remaining unemployed, or by comparing number of quarters vs number placed in jobs each quarter, or by comparing number of quarters vs total number placed in jobs at the end of each quarter. All graphs will be nonlinear.

9. Create a graph for the Hardy area church program over a five quarter period.

   A graph could be created by comparing number of quarters vs number of remaining unemployed, or by comparing number of quarters vs number placed in jobs each quarter, or by comparing number of quarters vs total number placed in jobs at the end of each quarter. All graphs will be linear.
   With Excel it is easy to place both graphs in one chart.

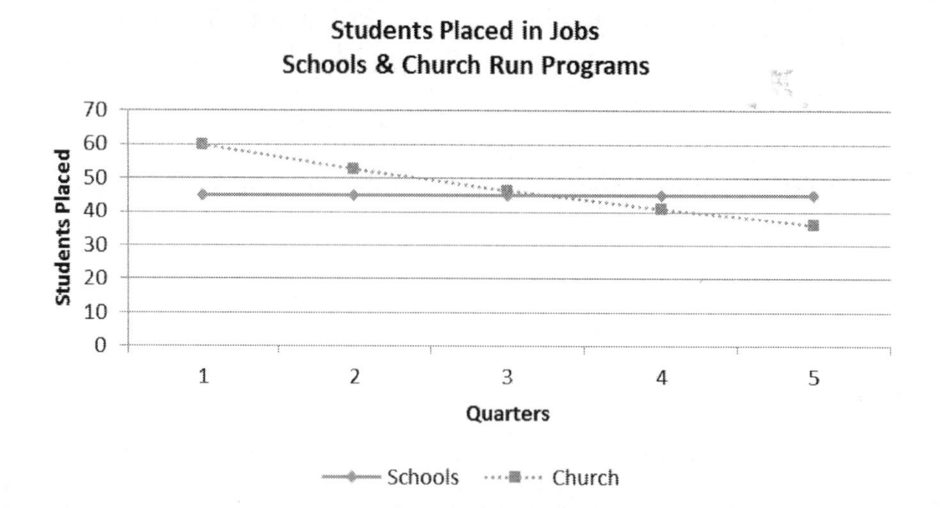
Students Placed in Jobs
Schools & Church Run Programs
Students Placed
70
60
50
40
30
20
10
0
1
2
3
4
5
Quarters
Schools
Church

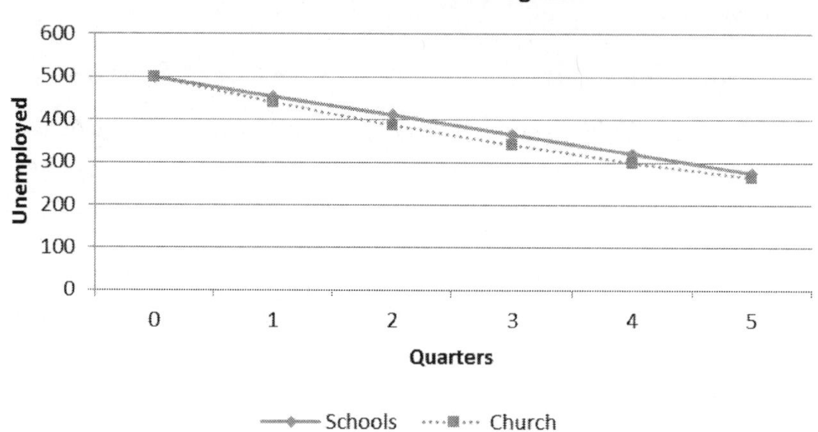
Students Still Unemployed
Schools & Church Run Programs
Unemployed
600
500
400
300
200
100
0
0
1
2
3
4
5
Quarters
Schools
Church

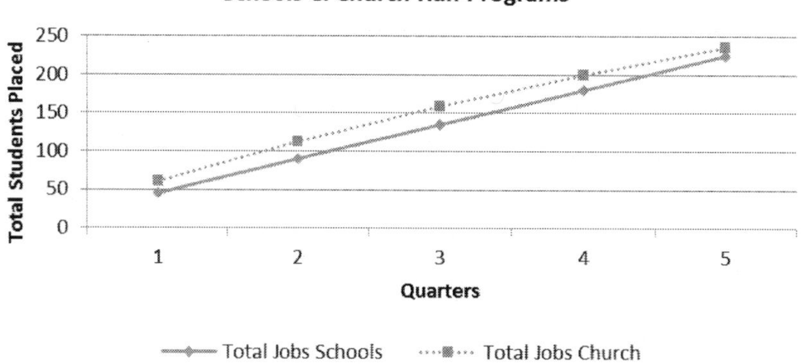
Total Students Placed in Jobs
Schools & Church Run Programs
Total Students Placed
250
200
150
100
50
0
1
2
3
4
5
Quarters
Total Jobs Schools
Total Jobs Church

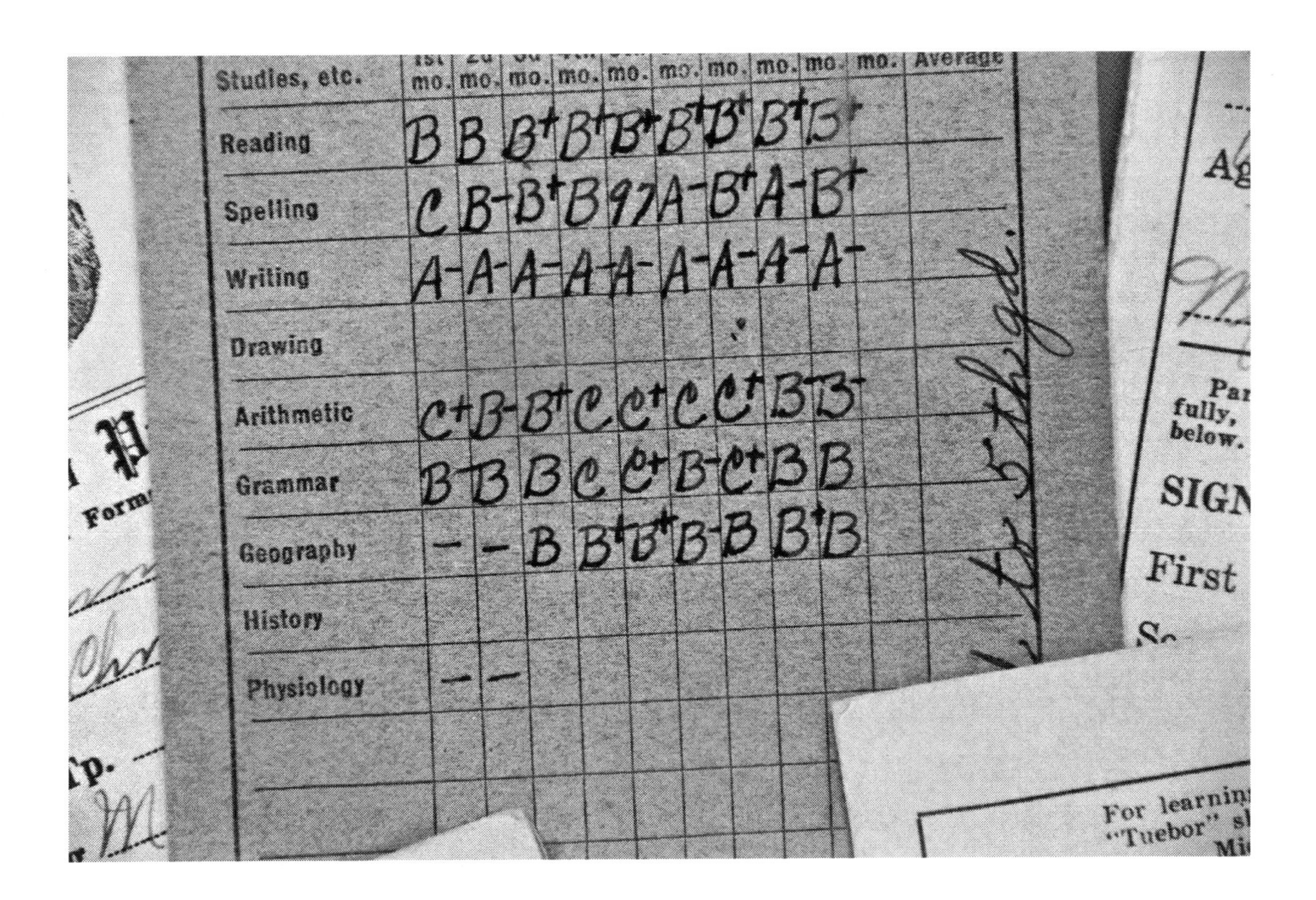

Activity 13:

Grades — Weighted Averages

Mathematical goals

The student will calculate the letter grade for different courses in which exams, projects, and reports are given different weights.

The student will:

- Read and use data presented in table format
- Use percentages to calculate weighted scores
- Set up algebraic expressions
- Solve algebraic equations

Before the lesson (5-10 minutes)

Put the paper and pencil down and practice some mental mathematics.

Number Talk Possibilities:

Select two or three depending on student abilities.

- $4 \times 12 + 8 \times 20 =$
- $2 \times 80 + 3 \times 10 =$
- 25% of 80 + 33 ⅓ % of 60 =
- 50% of 180 + 75% of 120 =

Weighted Averages

Denise Markham is nearing the end of her junior year at Bruce Kent High School. She has already received her grades for two courses, Math and English literature. The school assigns letter grades based on the score ranges in Table 1.

Score	Letter Grade
93-100	A
90-92.9	A-
87-89.9	B+
83-86.9	B
80-82.9	B-
77-79.9	C+
73-76.9	C
70-72.9	C-
67-69.9	D+
63-66.9	D
60-62.9	D-
0-59.9	F

Table 1: Score range and letter grade

The math grade is based simply on two exams, a midterm and final. The two exams are not given equal weight. The final exam counts for 60% of the grade. She studied extra hard and did much better on the final exam than on the midterm. She scored 83 on the midterm and 95 on the final.

	Math	
	Weight Factor	**Score**
Midterm	40%	83
Final	60%	95

Table 2: Math grades and weights

To find her total score in math, Denise multiplied each score by its weight factor.

$$0.40 \times 83 + 0.60 \times 95 = 90.2$$

She barely made it into the A- range.

1. What letter grade would she have earned if the two exams were equally weighted?

English literature was not Denise's strongest subject. The final grade was based on two exams and a final assignment. Her scores are recorded in Table 2.

	English Literature	
	Weight Factor	**Score**
Exam 1	30%	90
Exam 2	30%	84
Final Assignment	40%	83

Table 3: English literature grades and weights

Denise found her total score in English Literature by multiplying each score by its weight factor. This time there were three factors:

$$0.30 \times 90 + 0.30 \times 84 + 0.40 \times 83 = 85.4$$

Her English literature grade was a B.

Denise has two more major courses. Each of them has a big assignment that is due next week. She did not do well on the Science exam. She earned only a 78. However, she had a great idea for her Science project and was working hard to finish it. She was excited because it counts for 65% of the grade. She was wondering if there was any chance she could still earn an A- for this class if she did a whiz bang job. Her brother Darnell said they could use algebra to help figure out the lowest grade she needed to earn an A-.

	Science	
	Weight Factor	Score
Exam	35%	78
Project	65%	

Table 4: Science grades and weights

Darnell set up a mathematical expression to represent the weighted final score in Science.

Let y = the score on the project.

Then her total score in science = $0.35(78) + 0.65y$

If she only scored a 70 on the project, her total score would be

$0.35 \times 78 + 0.65 \times 70 = 72.8$

This would earn her only a C grade.

2. Use the mathematical expression to find her total score and letter grade for the project scores listed in Table 5.

	Science	
	0.35(78) + 0.65y	
Project Score y	Total Score	Letter Grade
60		
70	72.8	C
80		
90		
100		

Table 5: Total scores and letter grades for different scores on the project

3. Based on the total scores you found in Table 5, how could Denise earn an A- grade?

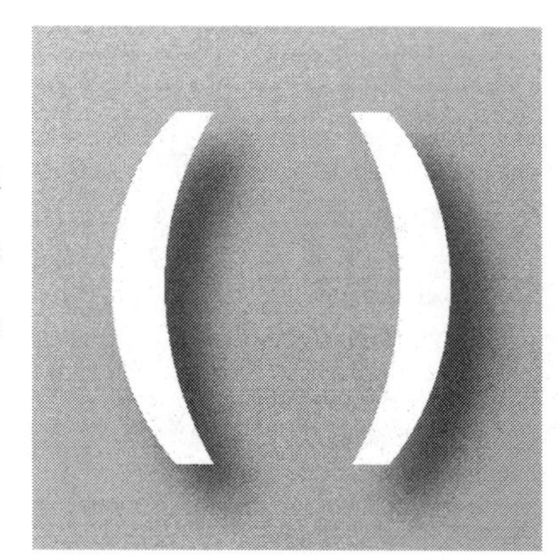

But Denise wanted to know the lowest score needed to earn an A- grade. Darnell then set the algebraic expression equal to the lowest score needed to earn an A- for the class. The lowest score for earning an A- is 90, so he wrote:

$0.35 \times 78 + 0.65y = 90$

"But we can use parentheses instead of a times sign," Denise remembered. "And we don't need any sign for multiplying y by a number!" So, they rewrote the equation like this:

$0.35(78) + 0.65y = 90$

$27.3 + 0.65y = 90$

Next, Darnell solved for y, the unknown value of the variable. First, he subtracted 27.3 from both sides of the equation to eliminate 27.3 from the left side. He subtracted, because subtraction is the inverse of addition.

$27.3 - 27.3 + 0.65y = 90 - 27.3$

$0.65y = 62.7$

Now only 0.65 times y remained on the left side of the equation. The right side still had just a number. To finish solving for y, Darnell divided both sides of the equation by 0.65, because division is the inverse of multiplication. That left only y on the left side of the equation:

$0.65y / 0.65 = 62.7 / 0.65$

$y = 96.46$

Now they knew she would earn an A- if she scored more than 96.46 on the final project. Since the teacher only gives whole number grades, they rounded up to 97.

Denise believed her project was great. However, she was not sure she had enough time to write a great final report. She wondered what score she needed to earn a B+.

4. What is the minimum score Denise needs to earn a B+ in Science?

Denise's final class is American History. Unfortunately, she did not do well on the exams. Her scores are given in Table 6.

	American History	
	Weight Factor	**Score**
Exam 1	25%	78
Exam 2	25%	81
Final Report	50%	

Table 6: American history grades and weights

Denise enjoys reading and believes she can do much better on the final report. Denise was wondering if there was any chance that she could still earn an A- for the course. She first needed to write a mathematical expression for finding her total score.

Let x = the score on final report

5. Write a mathematical expression to determine her weighted total score using x to represent her score on the final report.

6. What would her total score and letter grade be if she earned a perfect 100 on the final report?

Part 2 – Solve algebraic equation

Her brother, Darnell, told her be more realistic. He said she should try earning a B+ and no less than a B.

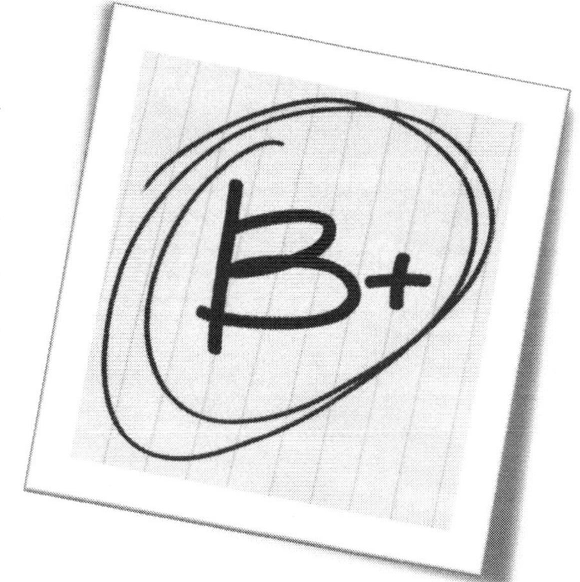

7. Set up an algebraic equation that can be used to find the minimum score she would need on the final report to earn a B+ in American History?

8. Solve the equation and find the minimum score she would need on the final report to earn a B+ in American History?

9. What is the minimum score she would need on the final report to earn a B in American History?

Project Idea:

Describe the grading method that your teachers use in each of your classes. What component of the grade is given the highest weight? What is given the smallest weight? If you have completed part of the class, determine what score you would need in the remainder to increase your score by five percentage points.

Practice problems

Leo Capri and Matt Demon are working on a science project for a state competition. They are looking for a third person. Leo and Matt want someone who can help with the writing, has strong data analysis skills, and can make a good presentation. Three potential teammates are listed in Table 7. They are Angelina Foley, Steve Wonk and Isaiah Newton. They use a scale of 1 to 4 to rate each person's skills. A 4 is the highest score and 1 is the lowest. No one candidate is the best in each category. Thus, there is no obvious best teammate.

Name	Writer	Analyst	Presenter
Angelina Foley	2	4	2
Steve Wonk	1	3	4
Isaiah Newton	4	2	1

Table 7: Candidate scores on different skills

To determine the best candidate, Leo and Matt decide to calculate a weighted score. Each category, writer, analyst and presenter will be assigned a percentage. The percentages will add up to 100%. However, Leo and Matt have different preferences. Leo wants someone very strong in data analysis. He assigns 50% to Analyst. Matt would like someone with good writing skills and gives it 50%. All the percentages are listed in Table 8.

	Writer	Analyst	Presenter
Leo Capri's preference	30%	50%	20%
Matt Demon's preference	50%	20%	30%

Table 8: Percentages assigned to different skills

1. Which candidate would Leo prefer?

2. Which candidate would Matt prefer?

3. Who do you recommend they choose and why?

Baseball

A weighted average is one in which different data in the data set are given different "weights". One example is the "Slugging Average" in baseball. A batter's slugging average (also called slugging percentage) is computed by:

SLG = ((1 × SI) + (2 × DO) + (3 × TR) + (4 × HR)) / AB

where: SLG = slugging percentage

SI = number of singles

DO = number of doubles

TR = number of triples

HR = number of home runs

AB = number of at-bats

The weight is the number of bases for each type of hit. Here, each single has a "weight" of one, each double has a "weight" of two, etc. An at-bat without a hit has a "weight" of zero.

4. Miguel Cabonara started the season with a hot hitting streak. He has had ten at-bats so far this season, which included 4-singles, 1-double, and 2-home runs. In the other three at-bats, he did not get a hit. Calculate Miguel's slugging average.

5. In the next game, Miguel had three at-bats and was hitless. Calculate his new slugging average.

6. Calculate Miguel's new slugging average, after his next at bat. Consider all possible outcomes, i.e., no hit, a single, a home run, etc.

7. Rocky Montana plays baseball and is off to a slow start. In his first ten at-bats, he only had

one single and one triple. How many consecutive home runs does Rocky need to hit to raise his slugging average to 1.00?

An individual's batting average is a common measure of performance. It is calculated by dividing the number of hits by the number of at bats. There are several types of at-bat that are not counted. For example, a walk and a sacrifice fly are not counted as an at-bat.

Batting Average = (number of hits)/(number of at-bats)

8. Calculate the slugging average and batting average for each player in Table 9. Which of the 5 players below would you select to be the Most Valuable Player of the team. Explain your thinking.

Name	Singles	Doubles	Triples	Home Runs	At Bats	Slugging Average	Batting Average
J.D. Martin	83	32	4	24	470		
Curtis Grander	47	41	12	37	504		
Justin Verygood	63	11	2	30	515		
Prince Field	78	23	1	38	522		
Brandon Inch	107	28	13	20	498		

Table 9: Slugging Percentage and batting average for five players

9. If the slugging average is "weighted" 60% and the batting average is "weighted" 40%, does that affect your MVP selection? Explain.

Activity 13: Teachers' guide

Thinking Through a Lesson Protocol

Standards:

6.EE.A.2.A: Write expressions that record operations with numbers and with letters standing for numbers.

6.EE.A.2.B: Identify parts of an expression using mathematical terms (sum, term, product, factor, quotient, coefficient); view one or more parts of an expression as a single entity.

6.EE.A.2.C: Evaluate expressions at specific values of their variables. Include expressions that arise from formulas used in real-world problems. Perform arithmetic operations, including those involving whole-number exponents, in the conventional order when there are no parentheses to specify a particular order (Order of Operations).

6.RP.A.3.C: Find a percent of a quantity as a rate per 100 (e.g., 30% of a quantity means 30/100 times the quantity); solve problems involving finding the whole, given a part and the percent.

7.RP.A.3: Use proportional relationships to solve multistep ratio and percent problems.

Mathematical Practices:

MP1: Make sense of problems and persevere in solving them.

MP2: Reason abstractly and quantitatively.

MP3: Construct viable arguments and critique the reasoning of others.

MP4: Model with mathematics.

Setting up the problem - Launch	
Selecting tasks/goal setting	(10 minutes) Have students discuss different grading schemes from different teachers and identify which they think might be weighted averages and why.
Questions	In high school, each semester grade counts as 40% and the final exam grade counts as 20% of the final grade in the course. Experiment with some fictitious grades and practice calculating the final grade. *Example: 40% of 82 + 40% of 76 + 20% of 94 = ...*

Monitoring student work - Explore		
Strategies and misconceptions-Anticipating	**Who - Selecting and sequencing**	**Questions and statements - Monitoring**
(10 Minutes) Read the text, look at the tables, and answer #1.		
(15 Minutes) Continue reading the text and complete the table in #2 (Table 5) and answer #3.		
(10 Minutes) Continue reading the text and answer #4.		
Continue reading the text and answer questions #5 and #6.		
Continue and answer #7-#9.		

Monitoring individual student work - Explore		
Strategies and misconceptions - Anticipating	**Who - Selecting and sequencing**	**Questions and statements - Monitoring**
For off-task students or for students that seem to be self-conscious about you listening to them share.		I am just listening or looking to find out how you are working on the problem. This helps me think about what we will do later.
For students that appear to be stuck. Also for when you are having a difficult time understanding their strategies.		Can you tell me a little about your reading? How would you describe the problem in your own words? What facts do you have? Could you try it with simpler numbers?
For students that want to ask you questions, these are ways to uncover their thinking and judge to what extent you want to respond.		Why are you interested in more information about that? Let me say a little about that part. Tell me what you've thought about so far. What do you know?

Managing the discussion - Summarize	
Parts of discussion - Connecting	**Questions and statements - Connecting**
Launching the discussion: Select the problems in questions #7-9 that students are struggling with or you wish to share out.	Will team 1 start us off by sharing one way of working on this problem? Please raise your hand when you are ready to share your solution. What did you do first when you were working on this problem? Let's start by clearing up a few things about the problem. Let's list some key parts in this problem. What was unclear in the problem?
Eliciting and uncovering student strategies	Joe would you be willing to start us off? What have you found so far? Can you repeat that? Can you explain how you got that answer? How do you know? Walk us through your steps. Where did you begin? Can you show us?
Focusing on mathematical ideas	Can you explain why this is true? Does this method always work? How is Bob's method similar to Kelly's method? What do all the solutions have in common? What would happen if I changed the numbers to _____?
Encouraging interactions	Do you agree or disagree with Kahlil's idea? What do others think? Would someone be willing to repeat what Tom just said? Would anyone be willing to add on to what Sue just said?
Concluding the discussion	Can anyone tell me some of the big ideas that we learned today? How would you explain what we learned today to a 5th grader? Some of the key points from our discussion today are . . . Tomorrow we will continue our exploration of ______ beginning with the idea from today that __________.
Post lesson notes	You may wish to assign the practice problems that you feel would benefit the students.

Solutions to text questions

1. What letter grade would she have earned if the two exams were equally weighted?

   $\frac{83+95}{2} = \frac{178}{2} = 89$

   This earns a B+.

2. Use the algebraic expression to find her total score and letter grade for the project scores listed in Table 5.

Project Score	Science 0.35(78) + 0.65y	
	Total Score	**Letter Grade**
60	*66.3*	*D*
70	72.8	C
80	*79.3*	*C+*
90	*85.8*	*B*
100	*92.3*	*A-*

Table 5: Total scores and letter grades for different scores on the project

3. Based on the total scores you found in Table 5, how could Denise earn an A- grade?

   She needs a score of 100 on the project.

4. What is the minimum score Denise needs to earn a B+ in Science?

   Use the equation we developed and simply set the right side of the equation equal to 87, the lowest grade to earn a B+.

   $0.35(78) + 0.65y = 87$

   $27.3 + 0.65y = 87$

   $0.65y = 87 - 27.3 = 59.7$

   $y = 59.7 / 0.65 = 91.8$

   She would need to score at least 92 to earn a B+ grade

5. Write a mathematical expression to determine her weighted total score using x to represent her score on the final report.

   $0.25(78) + 0.25(81) + 0.5x = \text{total score}$

   $39.8 + 0.5x = \text{total score}$

6. What would her total score and letter grade be if she earned a perfect 100 on the final report?

39.8 + 0.5x = total score

Substitute 100

39.8 + 0.5(100) = 89.7 This would earn a B+ grade

7. What is the minimum score she would need on the final report to earn a B+ in American History?

   0.25(78) + 0.25(81) + 0.5x = 87

   0.5x = 47.25

   x = 94.5, but if only whole number scores are given, then 95.

8. What is the minimum score she would need on the final report to earn a B in American History?

   0.25(78) + 0.25(81) + 0.5x = 83

   39.75 + 0.5x = 83

   0.5x = 43.25

   x = 43.25/0.

   x = 86.5, but since only whole number scores are given, then 87 is the answer.

Solutions to Practice Problems

Leo Capri and Matt Demon are working on a science project for a state competition. They are looking for a third person. Leo and Matt want someone who can help with the writing, has strong data analysis skills, and can make a good presentation. Three potential teammates are listed in Table 7. They are Angelina Foley, Steve Wonk and Isaiah Newton. They use a scale of 1 to 4 to rate each person's skills. A 4 is the highest score and 1 is the lowest. No one candidate is the best in each category. Thus, there is no obvious best teammate.

Name	Writer	Analyst	Presenter
Angelina Foley	2	4	2
Steve Wonk	1	3	4
Isaiah Newton	4	2	1

Table 7: Candidate scores on different skills

To determine the best candidate, Leo and Matt decide to calculate a weighted score. Each category, writer, analyst and presenter will be assigned a percentage. The percentages will add up to 100%. However, Leo and Matt have different preferences. Leo wants someone very strong in data analysis. He assigns 50% to Analyst. Matt would like someone with really good writing skills and gives it 50%. All the percentages are listed in Table 8.

	Writer	Analyst	Presenter
Leo Capri's preference	30%	50%	20%
Matt Demon's preference	50%	20%	30%

Table 8: Percentages assigned to different skills

1. Which candidate would Leo prefer?

Leo's Preference	*Writer*	*Analyst*	*Presenter*	*TOTAL*
Angelina Foley	$2\times0.30=0.6$	$4\times0.50=2$	$2\times0.20=0.4$	*3*
Steve Wonk	$1\times0.30=0.3$	$3\times0.50=1.5$	$4\times0.20=0.8$	*2.6*
Isaiah Newton	$4\times0.30=1.2$	$2\times0.50=1$	$1\times0.20=0.2$	*2.4*

Based on the weighted points of Leo's preferences, Leo would prefer Angelina.

2. Which candidate would Matt prefer?

Matt's Preference	***Writer***	***Analyst***	***Presenter***	***TOTAL***
Angelina Foley	$2\times0.50 = 1$	$4\times0.20 = 0.8$	$2\times0.3 = 0.6$	*2.4*
Steve Wonk	$1\times0.50 = 0.5$	$3\times0.20 = 0.6$	$4\times0.3 = 1.2$	*3.3*
Isaiah Newton	$4\times0.50 = 2$	$2\times0.20 = 0.4$	$1\times0.3 = 0.3$	*2.7*

Based on the weighted points of Matt's preferences, Matt would prefer Steve.

3. Who do you recommend they choose and why?

Based on the weighted combined points of Leo and Matt, the total scores would be: Angelina, $3 + 2.4 = 5.4$, *Steve,* $3.3 + 2.6 = 5.9$, *and Isaiah,* $2.4 + 2.7 = 5.1$, *the selection would be Steve.*

If the weight is changed in any of the categories, answers will vary.

Baseball

A weighted average is one in which different data in the data set are given different "weights". One example is the "Slugging Average" in baseball. A batter's slugging average (also called slugging percentage) is computed by:

SLG = ((1 × SI) + (2 × DO) + (3 × TR) + (4 × HR)) / AB

where: SLG = slugging percentage

SI = number of singles

DO = number of doubles

TR = number of triples

HR = number of home runs

AB = number of at-bats

Here, each single has a "weight" of one, each double has a "weight" of two, etc. An at-bat without a hit has a "weight" of zero.

4. Miguel Cabonara started the season with a hot hitting streak. He has had ten at-bats so far this season, which included 4-singles, 1-double, and 2-home runs. In the other three at-bats, he did not get a hit. Calculate Miguel's slugging average.

   SLG = (1×4 + 2×1 + 2×4) / 10
   SLG = 14 / 10
   SLG = 1.4

5. In the next game, Miguel had three at-bats and was hitless. Calculate his new slugging average.

   SLG = (14 + 0) / 13
   SLG = 1.07

6. Calculate Miguel's new slugging average, after his next at bat. Consider all possible outcomes, i.e., no hit, a single, a home run, etc.

   No Hit: (14 + 0) / 14 = 1.00
   Single: (14 + 1) / 14 = 1.07
   Double (14 + 2) / 14 = 1.14
   Triple (14 + 3) / 14 = 1.21
   Home Run (14 + 4) / 14 = 1.28

7. Rocky Montana plays baseball and is off to a slow start. In his first ten at-bats, he only had

one single and one triple. How many consecutive home runs does Rocky need to hit to raise his slugging average to 1.00?

$(1 \times 1 + 1 \times 3) / 10 = 0.4$
$(1 \times 1 + 1 \times 3 + 1 \times 4) / 11 = 0.727$ (one home run)
$(1 \times 1 + 1 \times 3 + 2 \times 4) / 12 = 1.00$ (two home runs)
Two more at bats with two consecutive home runs will bring his slugging average up to 1.00.

8. Based on calculating the slugging average and batting average, which of the 5 players below would you select to be the Most Valuable Player of the team. Explain your thinking.

Batting Average = (number of hits) / (number of at-bats)

Name	Singles	Doubles	Triples	Home Runs	At Bats	Slugging Average	Batting Average
J.D. Martin	83	32	4	24	470	*0.543*	*0.304*
Curtis Grander	47	41	12	37	504	*0.621*	*0.272*
Justin Verygood	63	11	2	30	515	*0.410*	*0.206*
Prince Field	78	23	1	38	522	*0.534*	*0.268*
Brandon Inch	107	28	13	20	498	*0.566*	*0.337*

Table 9: Slugging Percentage and batting average for five players

Answer will vary. If you combine the two scores or average them, Brandon Inch has the highest total, 0.903. The average of the two measures is 0.452. This gives the two measures equal weight. The second best is Curtis Grander. His total is 0.893. The average is 0.447.

9. If the slugging average is "weighted" at 60% and the batting average is "weighted" at 40%, does that affect your MVP selection? Explain.

Brandon Inch $(0.6 \times 0.621 + 0.4 \times 0.272) = 0.481$
Curtis Grander $(0.6 \times 0.566 + 0.4 \times 0.337) = 0.474$

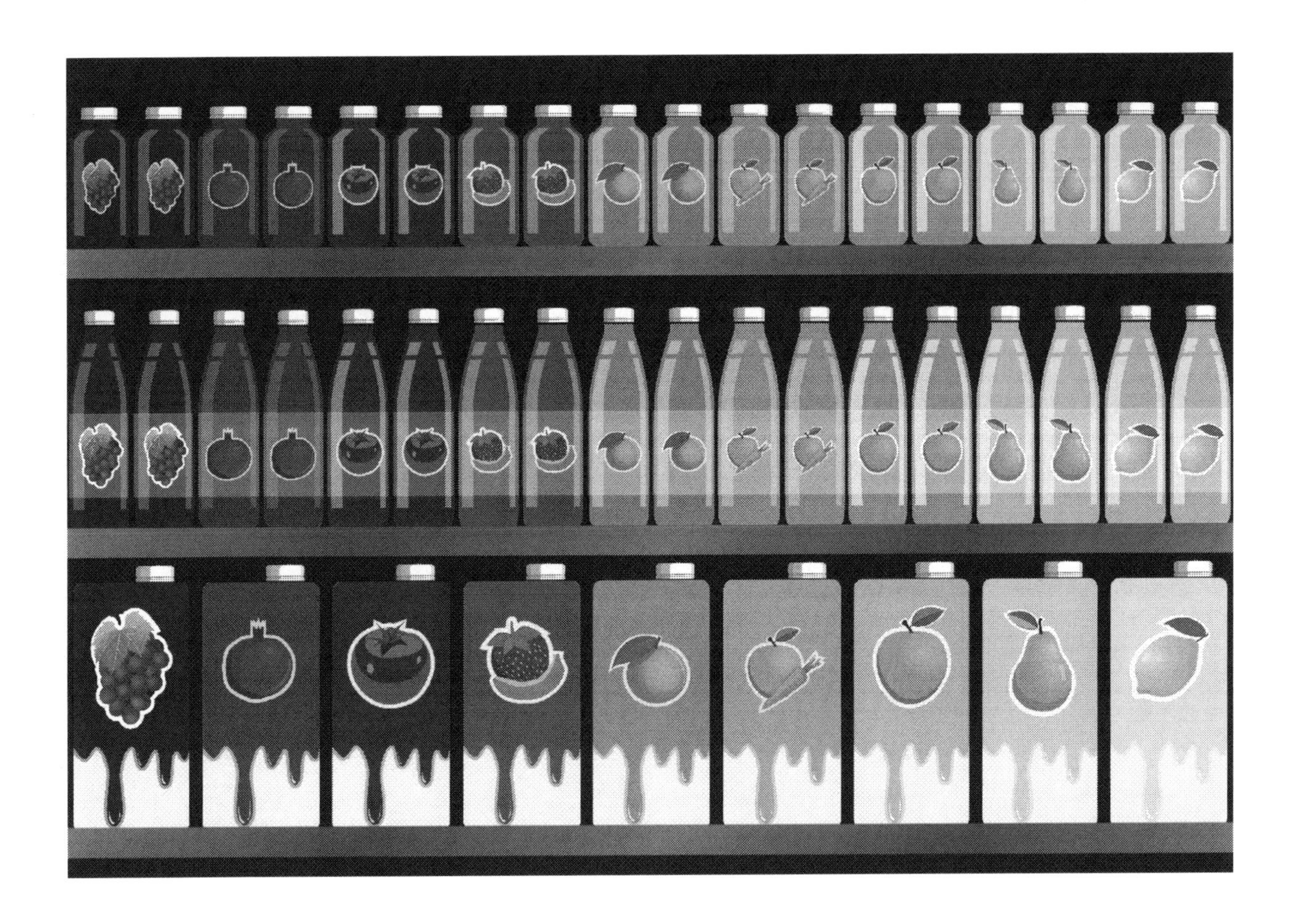

Activity 14:
Open24 Slushy Sales

Mathematical goals

The student will calculate totals of several measures. This example involves comparing compounding percentages with fixed rate of decrease.

The student will:

- Read and use data presented in table format
- Weight values by percentages to calculate totals
- Perform financial analysis
- Interpret a pie chart

Before the lesson (5-10 minutes)

Put the paper and pencil down and practice some mental mathematics.

Number Talk Possibilities:

Select two or three depending on student abilities.

- 25% of 60 + 75% of 40 is ?
- 80% of 20 + 20% of 80 is ?
- 40% of 50 + 60% of 50 is ?
- 33 ⅓% of 60 + 33 ⅓% of 60 + 33 ⅓ % of 60 is ?

Open24 Slushy Sales

Marjorie Star is the district manager for Open24, a chain of convenience stores. Open24 stores are located at fuel stops along major highways. They are open 24 hours a day. They sell a lot of a frozen drink called Slushy. Slushies are sold in three sizes, medium (16 oz. cup), large (22 oz. cup) and X-large (28 oz. cup). Table 1 has data about all three sizes. This includes the price, the profit and the percent of sales. The 16 oz. Slushy sells for $1.50. The store makes a profit of $0.75 on each sale of the 16-oz. size. Usually, the 16-oz. size is the most popular. Fifty percent of the customers choose the 16-oz. size.

Size (ounces)	Price	Profit	Percent of sales
16	$1.50	$0.75	50%
22	$1.80	$1.00	36%
28	$2.10	$1.25	14%

Table 1: Slushy price, profit, and sales

Open24 has a store at the I-94 exit near Portage, MI. This store sells an average of 750 Slushies each day.

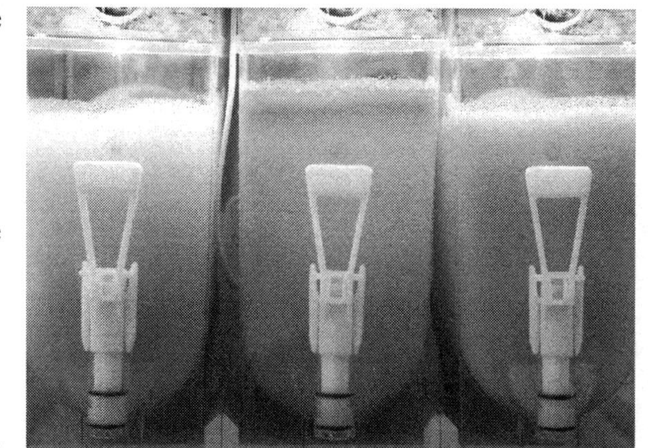

1. How many of each size of Slushy does the Portage store sell per day?

2. How many ounces of Slushy does the Portage store sell per day?

3. How much money does the store take in from Slushy sales per day?

4. How much profit does the store earn from Slushy sales per day?

5. On average, how much profit does the store make for each Slushy?

All gas pumps at this location have High Octane Television (HOTV) running non-stop. HOTV sells ads along with its stream of news and personal interest stories. Adrian Prime is a statistical analyst for Open24. Marjorie asked Adrian to analyze data from other states about the value of advertising. Open24 has two different Slushy ad campaigns. One set of 15 second ads costs $28 per day. Adrian's analysis found buying these 15 second ads could increase Slushy sales by 8%.

6. Would you recommend spending $28 per day on these 15 second ads? Justify your answer.

Other ads try to get customers to buy larger sizes that make more profit. These ads run for 20 seconds each. The cost to run these ads is $35 per day. Adrian analyzed data from many ad campaigns that Open24 has run on HOTV. The 20 second ads can change the percentage of customers who buy different sizes. Based on these studies, Adrian reported the new percentages in Table 2. Figure 1 shows the comparison of current and with advertising.

Size	Percent of sales	
	Current	With 20 second Ads
16	50%	25%
22	36%	51%
28	14%	24%

Table 2: Size percentages before and after advertising

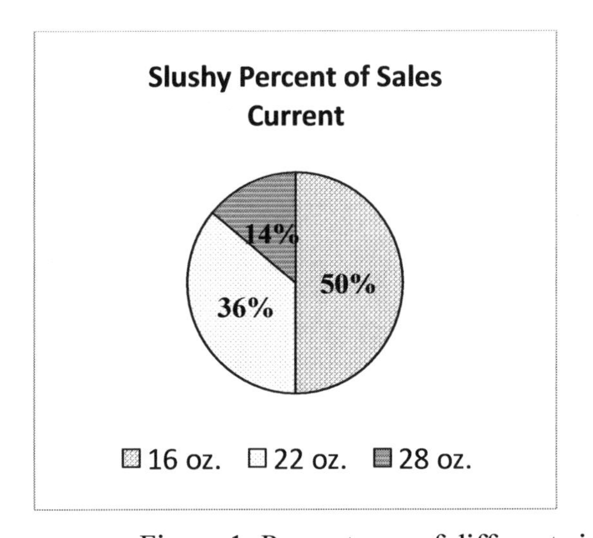

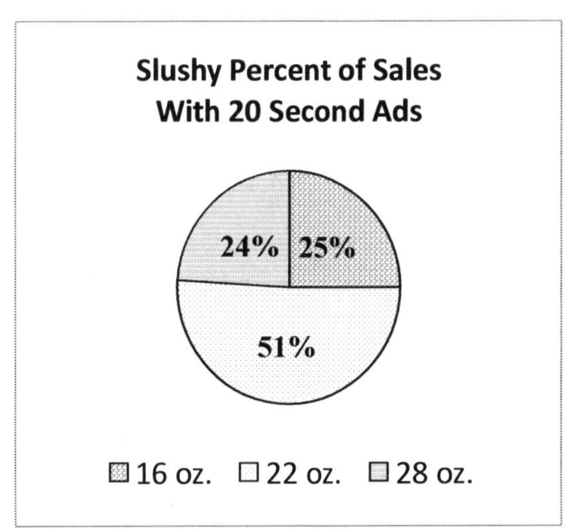

Figure 1: Percentages of different sized Slushies: Current and With Ads

7. What are the biggest differences between the two pie charts?

8. Would you recommend spending $35 per day to increase the purchases of the larger Slushy? Justify your answer.

9. Which of the two ad campaigns would you recommend using? Justify your answer.

Project Idea:

With permission of the store manager, collect data about people coming into a 7-11 and buying a Slurpee. How many purchase small, medium, large and XL? What percentage of customers take a single flavor and what percentage mix flavors? What is the most popular flavor?

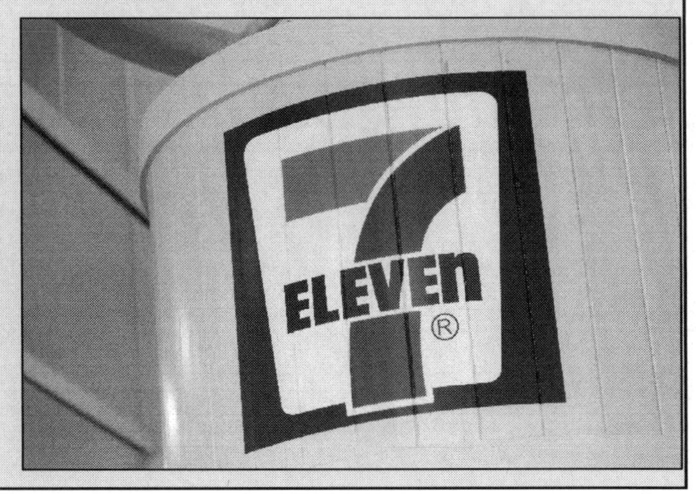

Practice problems

The large multiplex movie theatre sells popcorn in three different sizes; small, medium and large. Table 3 shows the price and profit on the three different sizes. The sales percent are based on data collected from last month.

Size	Price	Profit	Percent of sales
Small	$3.50	$1.75	8%
Medium	$4.75	$2.50	60%
Large	$6.00	$3.75	32%

Table 3: Price and profit for different sizes

On a typical weekend, they sell 900 containers of popcorn per day.

1. On average, how many of each size of popcorn will the local theater sell per day on a typical weekend?

2. On average, how much money will the theater take in each weekend day from popcorn sales per day?

3. On average, how much profit will the theater earn from popcorn sales each day?

On a busy weekend with exciting new releases, they can sell 1200 containers of popcorn per day.

4. How many of each size of popcorn will the local theater sell per day on a busy weekend?

5. How much money will the theater take in each day from popcorn sales per day?

6. How much profit will the theater earn from popcorn sales each day?

7. Create a circle graph that shows the "Percent of Sales" of each size of popcorn per day.

8. Twenty five percent of all daily sales of popcorn occur during the matinee time period (before 4:00 pm). The remainder of sales occur after 4:00pm. Find the number of small, medium, and large popcorn sales during the matinee on a busy weekend.

The profit margin is determined by dividing the profit by the sale price and reported as a percent of the sales price.

9. What is the profit margin on each sized container of popcorn?

10. Which size has the greatest profit margin and why do you think that is?

Activity 14: Teachers' guide

Thinking Through a Lesson Protocol

Standards:

6.RP.A.3.C: Find a percent of a quantity as a rate per 100 (e.g., 30% of a quantity means 30/100 times the quantity); solve problems involving finding the whole, given a part and the percent.

7.RP.A.3: Use proportional relationships to solve multistep ratio and percent problems.

Mathematical Practices:

MP1: Make sense of problems and persevere in solving them.

MP2: Reason abstractly and quantitatively.

MP3: Construct viable arguments and critique the reasoning of others.

MP4: Model with mathematics.

Setting up the problem - Launch	
Selecting tasks/goal setting	(10 minutes) Have students discuss where they get soft drinks or slurpees in their neighborhoods. What sizes and flavors are available?
Questions	If you were to set up a statistical experiment and wanted to collect data at your local convenience store, how might you do it? What are the different data that can be collected? How might you organize it? How might you record it?

Monitoring student work - Explore		
Strategies and misconceptions- Anticipating	**Who - Selecting and sequencing**	**Questions and statements - Monitoring**
(15 Minutes) Read the text, look at the tables, and answer #1 thru #5.		
(15 Minutes) Continue reading the text and answer #6. Share student responses.		
(15 Minutes) Continue reading the text and answer #7 thru #9.		

Monitoring individual student work - Explore		
Strategies and misconceptions - Anticipating	**Who - Selecting and sequencing**	**Questions and statements - Monitoring**
For off-task students or for students that seem to be self-conscious about you listening to them share.		I am just listening or looking to find out how you are working on the problem. This helps me think about what we will do later.
For students that appear to be stuck. Also for when you are having a difficult time understanding their strategies.		Can you tell me a little about your reading? How would you describe the problem in your own words? What facts do you have? Could you try it with simpler numbers?
For students that want to ask you questions, these are ways to uncover their thinking and judge to what extent you want to respond.		Why are you interested in more information about that? Let me say a little about that part. Tell me what you've thought about so far. What do you know?

Managing the discussion - Summarize	
Parts of discussion - Connecting	**Questions and statements - Connecting**
Launching the discussion: Select the problems in questions #7-9 that students are struggling with or you wish to share out.	Will team 1 start us off by sharing one way of working on this problem? Please raise your hand when you are ready to share your solution. What did you do first when you were working on this problem? Let's start by clearing up a few things about the problem. Let's list some key parts in this problem. What was unclear in the problem?
Eliciting and uncovering student strategies	Joe would you be willing to start us off? What have you found so far? Can you repeat that? Can you explain how you got that answer? How do you know? Walk us through your steps. Where did you begin? Can you show us?
Focusing on mathematical ideas	Can you explain why this is true? Does this method always work? How is Bob's method similar to Kelly's method? What do all the solutions have in common? What would happen if I changed the numbers to _____?
Encouraging interactions	Do you agree or disagree with Kahlil's idea? What do others think? Would someone be willing to repeat what Tom just said? Would anyone be willing to add on to what Sue just said?
Concluding the discussion	Can anyone tell me some of the big ideas that we learned today? How would you explain what we learned today to a 5th grader? Some of the key points from our discussion today are . . . Tomorrow we will continue our exploration of ______ beginning with the idea from today that __________.
Post lesson notes	You may wish to assign the practice problems that you feel would benefit the students.

Solutions to text questions

1. How many of each size of Slushy does the Portage store sell per day? The number of 16 ounce Slushies is $0.5 \times 750 = 375$. The number of 22 ounce Slushies is $0.36 \times 750 = 270$.

   The number of 28 ounce Slushies is $0.14 \times 750 = 105$.

2. How many ounces of Slushy does the Portage store sell each day?

   Multiply the number of each size by the corresponding number of ounces to obtain the total number of ounces.

   $(0.50 \times 750 \times 16) + (0.36 \times 750 \times 22) + (0.14 \times 750 \times 28) = 14{,}880$ *oz.*

3. How much money does the store take in from Slushy sales per day?

   The logic is the same as above. Except we multiply each amount by the price of the corresponding Slushy.

   $(0.50 \times 750 \times \$1.50) + (0.36 \times 750 \times \$1.80) + (0.14 \times 750 \times \$2.10) = \$1{,}269.00$

4. How much profit does the store earn from Slushy sales per day?

   The logic is the same as above. Except we multiply each amount by the profit of the corresponding Slushy.

   $(0.50 \times 750 \times \$0.75) + (0.36 \times 750 \times \$1.00) + (0.14 \times 750 \times \$1.25) = \$682.50$

5. On average, how much profit does the store make for each Slushy?

   Total profit / number of Slushies

   $\$682.50 / 750 = 0.91$

   Alternatively, we can calculate the weighted sum. Fifty percent of the Slushies earn \$0.75, 36% earn \$1.00 and 14% earn \$1.25

   $(0.50 \times \$0.75) + (0.36 \times \$1.00) + (0.14 \times \$1.25) = \$0.91$

6. Would you recommend spending $28 per day on these 15 second ads? Justify your answer.

   Yes; The 8% increase in sales add 60 to the total.

   $750 + (0.08 \times 750) = 810$

   $(0.50 \times 810 \times \$0.75) + (0.36 \times 810 \times \$1.00) + (0.14 \times 810 \times \$1.25) = \$737.10$

Net increase in profit is.

$737.10 – $682.50 = $54.60

$54.60 – $28.00 = $26.60

The profit would increase by $26.60 per day.

Alternatively, they will sell 60 more Slushies.

The average profit per Slushy is $0.91

Extra profit = 60 × $0.91 = $54.60 which is more than the cost of advertising.

7. What are the major differences between the two pie charts?

   Sales of the 16 oz. Slushy decreased from 50% to 25%, but sales of the 22 oz. size increased from 36% to 51% and the sales of the 28 oz. size increased from 14% to 24%.

8. Would you recommend spending $35 per day to increase the purchases of the larger Slushy? Justify your answer.

   Yes;

   (0.25 × 750 × $0.75) + (0.51 × 750 × $1.00) + (0.24 × 750 × $1.25) = $748.125 ≈ $748.13

   $748.13 – $682.50 = $65.63

   $65.63 – $35.00 = $30.63

   The profit would increase by $30.63 per day.

9. Which of the two ad campaigns would you recommend using? Justify your answer.

   The second plan, because it would increase daily profit by $30.62 which is more than the $26.60 extra profit per day with the first plan. This is $4.02 more than the first plan.

Solutions to Practice Problems

Size	Price	Profit	Percent of sales	*Sell 900*		
				Sales	***Revenue***	***Profit***
Small	$3.50	$1.75	8%	*72*	*$252*	*$126*
Medium	$4.75	$2.50	60%	*540*	*$2,565*	*$1,350*
Large	$6.00	$3.75	32%	*288*	*$1,728*	*$1,080*
Totals				*900*	*$4,545*	*$2,556*

Table 3: Price and profit for different sizes

On a typical weekend, they sell 900 containers of popcorn per day.

1. How many of each size of popcorn will the local theater sell per day on a typical weekend?

   See Table above.

2. How much money will the theater take in each weekend day from popcorn sales per day?

   $4,545

3. How much profit will the theater earn from popcorn sales each day?

   $2,556

On a busy weekend with exciting new releases, they can sell 1200 containers of popcorn per day.

Size	Price	Profit	Percent of sales	*Sell 1200*		
				Sales	***Revenue***	***Profit***
small	$3.50	$1.75	8%	*96*	*$336*	*$168*
medium	$4.75	$2.50	60%	*720*	*$3,420*	*$1,800*
large	$6.00	$3.75	32%	*384*	*$2,304*	*$1,440*
Totals				*1200*	*$6,060*	*$3,408*

4. How many of each size of popcorn will the local theater sell per day on a busy weekend?

   See Table above.

5. How much money will the theater take in each day from popcorn sales per day?

   $6,060

6. How much profit will the theater earn from popcorn sales each day?

   $3,408

7. Create a circle graph that shows the "Percent of Sales" of each size of popcorn per day.

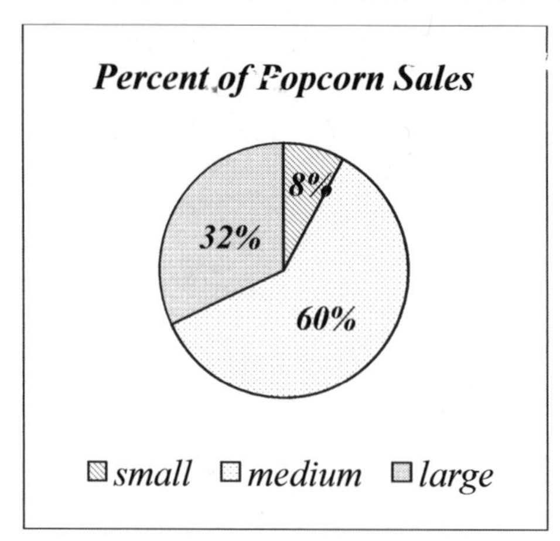

8. Twenty five percent of all daily sales of popcorn occur during the matinee time period (before 4:00pm). The remainder of sales occur after 4:00pm. Find the number of small, medium, and large popcorn sales during the matinee on busy weekend.

0.25(1200) = 300

The sales of small is 0.08(300) = 24

The sales of medium is 0.60(300) = 180

The sales of large is 0.32(300) = 96. OR 0.25(96) = 24

0.25(720) = 180

0.25(384) = 96

9. What is the profit margin on each sized container of popcorn?

See Table.

Size	Price	Profit	*Profit Margin*
Small	\$3.50	\$1.75	*1.75 / 3.50 = 0.50 ➔ 50%*
Medium	\$4.75	\$2.50	*2.50 / 4.75 = 0.53 ➔ 53%*
Large	\$6.00	\$3.75	*3.75 / 6.00 = 0.63 ➔ 63%*

10. Which size has the greatest profit margin and why do you think that is?

Answers will vary. The largest popcorn is 70% more expensive than the smallest. The amount of labor is takes to make popcorn, fill the largest container and collect payment is not that much more than for the smallest container. As a result, the cost is not proportional to the price and the profit margin is greater.

Activity 15:
Constructing Congressional Districts

Mathematical goals

The student will use past voting data to project the outcome of an upcoming election and analyze various redistricting plans. This example involves determining the redistricting plans and their corresponding percentages and the impact they might have on an election.

The student will:

- Read and use data presented in table format
- Analyze and reorganize population data
- Calculate percentages
- Extract data from regional zone maps

Before the Lesson (5-10 minutes)

Put the paper and pencil down and practice some mental mathematics.

Number Talk Possibilities:

Select two or three depending on student abilities.

- 53 is what percent of 100?
- 530 is what percent of 1000?
- 5300 is what percent of 10,000?
- 53,000 is what percent of 100,000?
- 30 is what percent of 300?
- 150 is what percent of 300?
- 180 is what percent of 300?

Create Congressional Districts

The US Constitution requires that the federal government conduct a census every ten years. A census is a count of all the people living in the US. It is used to decide how many of the 435 seats in the House of Representatives will go to each of the 50 states. The actual number of seats has not changed since 1913 even though it was not codified into law until the Reapportionment Act of 1929. This number could be changed by an act of any Congress, because it is not stated in the Constitution or any constitutional amendment. Each state decides how to construct its districts. There are only a few rules that limit how districts are constructed. Each district should have approximately the same population. The geographic pieces in a district should touch one another although a body of water could split the district. In most states the process of designing the districts is political. The governor and the state legislature must reach an agreement.

In eight states the population voted an amendment to the state constitution. These amendments require that redistricting be done by a group chosen from both major political parties. Seven states use an independent commission whose design is final. Another state uses non-partisan legislative staff to develop the plan. They are not allowed to use any political or election data to develop their plan. The government in that state can accept or reject the plan, but it cannot change the plan.

In each election for the House of Representatives, the candidate who receives the largest percentage of votes within the district wins the seat. If there are three or more candidates, the winner could receive much less than 50% of the vote. There are big differences in the nature of the populations from one district to the next. In one district, as many as 75% of the citizens might vote for one party. In another fewer than 30% of the citizens might vote for that same party.

In the US we vote for 435 individual congressmen. The winning percentage will vary from district to district. As a result, the national vote percentages do not match the percentage of winners. For example, in the 2012 election, Democratic candidates for the House of Representatives obtained a total of 1.4 million votes more than their Republican opponents. This corresponded to 50.6% of the total votes for these two parties. However, Republicans won 53.8% of the seats. There were 234 Republicans elected and only 201 Democrats in the 2012 election. In the following example, we explore how changing the boundaries of an election district might affect the number of Democrats or Republicans elected.

In the 2012 election, there were a total of 117.9 million votes cast for Democratic and Republican candidates for Congress. This was approximately 38% of the total estimated US population of 314.1 million people. Figure 1 uses a funnel to show how over three hundred million people result in 117.9 million votes. Approximately, 93% of the people living in the US are citizens of the US. This excludes legal immigrants who have not yet become citizens and illegal immigrants who are not eligible to become citizens. Of this total, only 209.3 million are registered to vote. This excludes people below the age of 18 who cannot vote and other people who did not bother to register. Finally, a large percentage of registered voters do not bother to vote. The number of actual voters in 2012 was 37.5% of the total population of the U.S.

2012 US Population (estimated) = 314.1 million
2012 US Citizens (estimated) = 292.1 million
Registered Voters (estimated) = 209.3 million
Actual Voters = 117.9 million

Figure 1: Funnel from US population to Actual Voters

Imaginary Metropolitan (Metro) Area

We have created an imaginary metropolitan area, the Sundial, with a population of 3 million, and four congressional districts. See Figure 2. The population is evenly spread among 12 smaller zones. Each zone has a population of 250,000. In this area, an estimated 40% of the people vote in congressional elections. This voting percentage is slightly above the national average. Thus, the number of voters in each zone is 100,000.

The Congressional Redistricting Commission (CRC) is charged with creating the new congressional districts. They will combine three zones to form each district. There is data from prior years about the number of people in each zone who voted for Democratic and Republican congressmen in the last election. The percentage of Democratic and Republican voters differs from zone to zone.

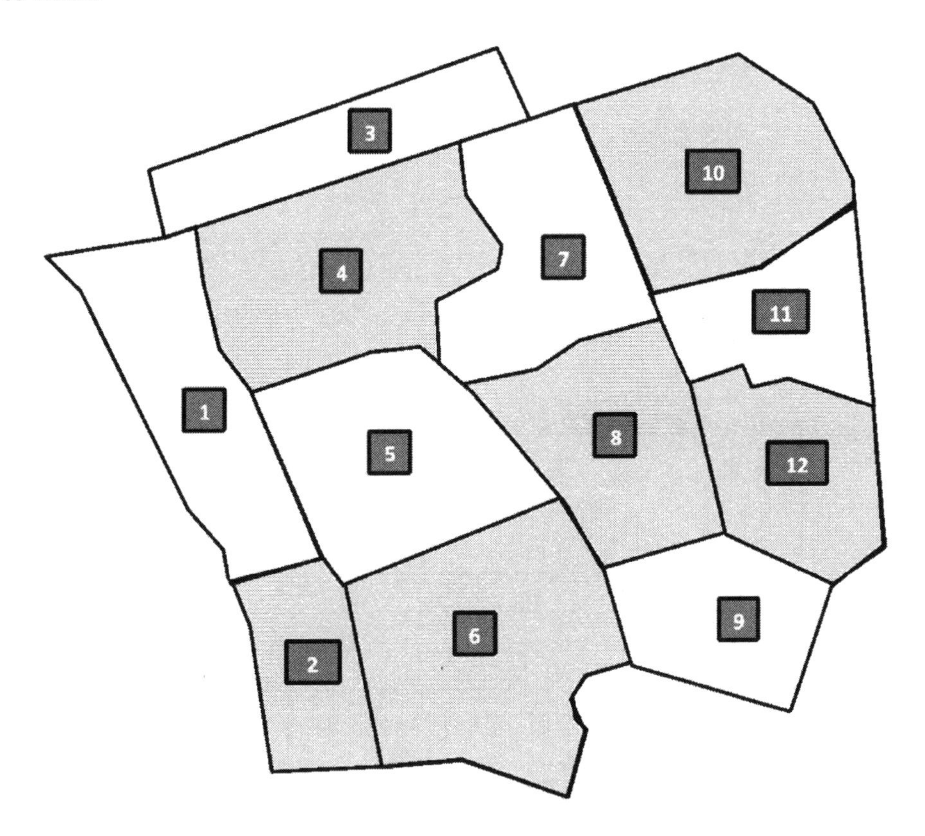

Figure 2: Map with Zone numbers

To simplify the visualization of new districts, the map with irregular boundaries was redrawn with rectangles. Figure 3 contains an equivalent map regarding the zones that are next to one another. (The zone sizes are not proportional.)

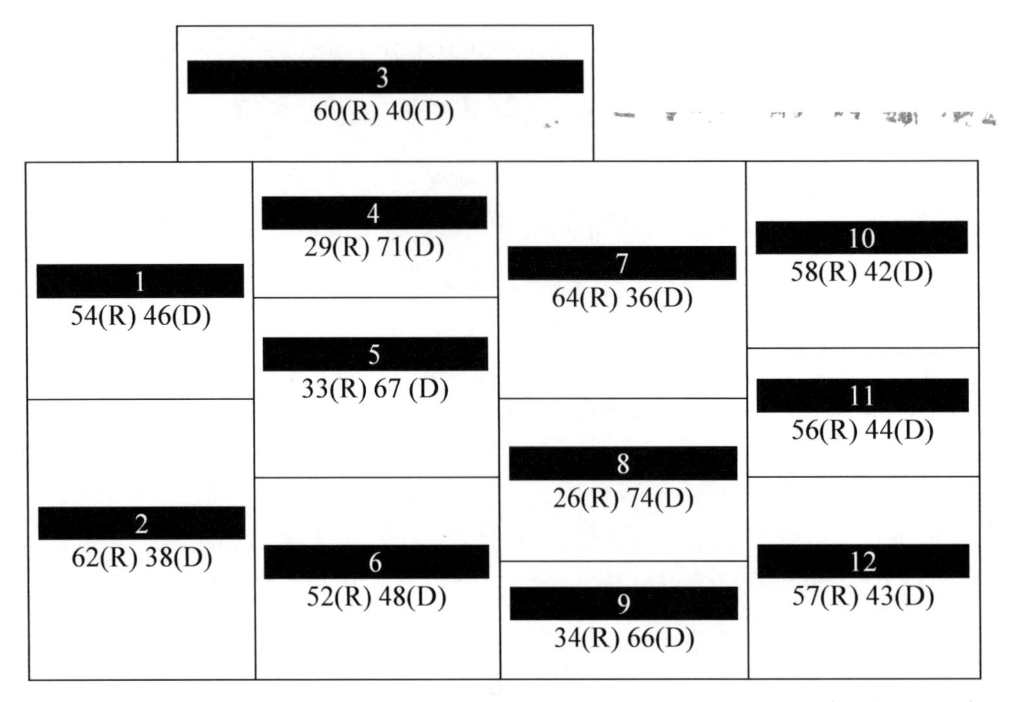

Figure 3: Zone numbers with Republican and Democratic voters in thousands

The maps are said to be equivalent if the rectangles maintain the same connections with and between zones as the actual map. For example, Zone 3, touches on zones 1, 4, and 7 in both figures and no other zones. Zone 5, touches many more zones. It touches zones 1, 2, 4, 6, 7 and 8. To prove that the two maps are equivalent, for each map, record the zones that are next to each of the zones listed in Table 1.

Zone	Zone Adjacency List	
	Actual Map	Rectangle Map
3	1, 4, and 7	1, 4, and 7
4	1, 2, 4, 6, 7 and 8	1, 2, 4, 6, 7 and 8
12		
7		
2		

Table 1: Neighboring zones

After the recent census, the congressional districts had to be redrawn. The way they are created can affect the number of Democrats and Republicans who win future elections. Figure 3 and Table 2 show the number of people from each party who live in each zone in the region.

Each zone has 100,000 people who voted in the last election. Below the zone number is the number of people who voted for Republicans and Democrats in that zone in the last election just before the census. Because each zone's voting population is 100,000, the number of Democratic and Republican voters can easily be converted into percentages. For example, Zone 4 is part of the inner city. A total of 71,000 people voted Democratic. This is equal to 71% of the 100,000 voters. In contrast in Zone 7, only 36,000 voted for Democrats, which is 36% of the total. In this metro area, a total of 615,000 votes were cast for Democratic candidates. This is 51.3% of the total of 1.2 million votes cast. A total of 585,000 were cast for Republicans, 48.7% of the total.

Zone	Republican (thousands)	Democrats (thousands)
1	54	46
2	62	38
3	60	40
4	29	71
5	33	67
6	52	48

Zone	Republican (thousands)	Democrats (thousands)
7	64	36
8	26	74
9	34	66
10	58	42
11	56	44
12	57	43

Table 2: Zone numbers with Republican and Democratic voters

Democratic Legislature Proposal – D

The Democratic Legislature set up a committee to construct the congressional districts. The committee recommended the plan presented in Figure 3. Consider what would happen if the area was divided simply into four quadrants. Each district would have a total of 300,000 votes. Any candidate who receives more than 150,000 votes wins. The upper left zone consists of Zone 1 with 54,000 Republican voters, Zone 3 with 60,000 Republican voters and Zone 4 with 29,000 Republican voters. This is a total of 143,000 Republicans out of 300,000 votes. The Democratic candidate would win this district with 157,000 votes if the voting matched the previous election totals. Democrats would win with 52.3% of the vote total.

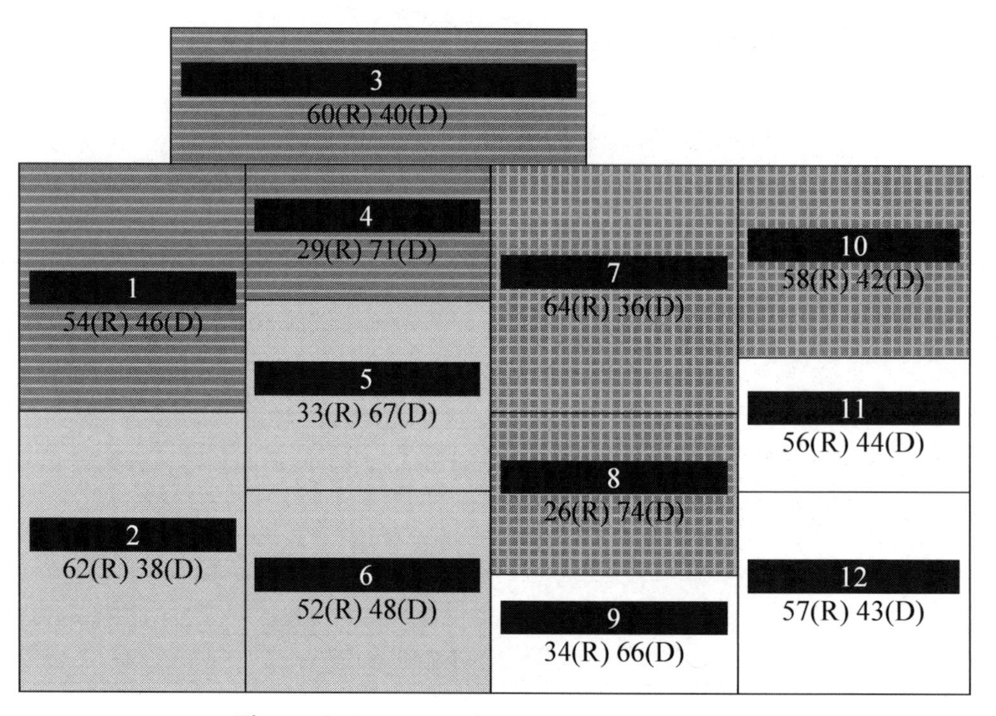

Figure 3: Democratic Legislature - plan D

District Location	Zone	Republican (000s)	Democrats (000s)	District Location	Zone	Republican (000s)	Democrats (000s)
Upper Left	1	54	46	Upper Right	7	64	36
	3	60	40		8	26	74
	4	29	71		10	58	42
Lower Left	2	62	38	Lower Right	9	34	66
	5	33	67		11	56	44
	6	52	48		12	57	43

Table 3: Democratic Legislature - plan D

District Location	Zone	Republican Totals and %	Democrats Totals and %	District Location	Zone	Republican Totals and %	Democrats Totals and %
Upper Left	1	143,000 47.7%	157,000 52.3%	Upper Right	7	148,000 49.3%	152,000 50.7%
	3				8		
	4				10		
Lower Left	2	147,000 49%	153,000 51%	Lower Right	9	147,000 49%	153,000 51%
	5				11		
	6				12		

Table 4: Democratic Legislature - plan D – totals by district

The upper right district consists of Zone 7 with 64,000 Republican voters, Zone 10 with 58,000 Republican voters and Zone 8 with 26,000 Republican voters. This is a total of 148,000 Republicans out of 300,000 votes. This would be a close election. The Democratic candidate would win this district with 152,000 votes if the voting matched the previous election totals. In the upper left district, Democrats received 52.3% of the vote in the last election. Table 4 summarizes the total number of votes for each party. It also presents the percentages.

1. Determine the totals and percentages for each of the two districts in the lower part of the map. Record these values in Table 4

2. Which party would likely win each of these districts?

3. Under this plan, how many congressional districts would likely be won by Democratic candidates?

4. Which district would likely have the closest election?

Republican Governor Proposal - R

Gerry Manderbilt, the Republican governor, can veto the final plan. His team has come up with a different idea about the districts. Three lower center zones (5, 8 and 9) in the city are combined to create one district. Most likely, a Democrat would easily win that district with 207,000 votes. This represents 69% of the total. The other districts are shown in Figure 4 and Table 5.

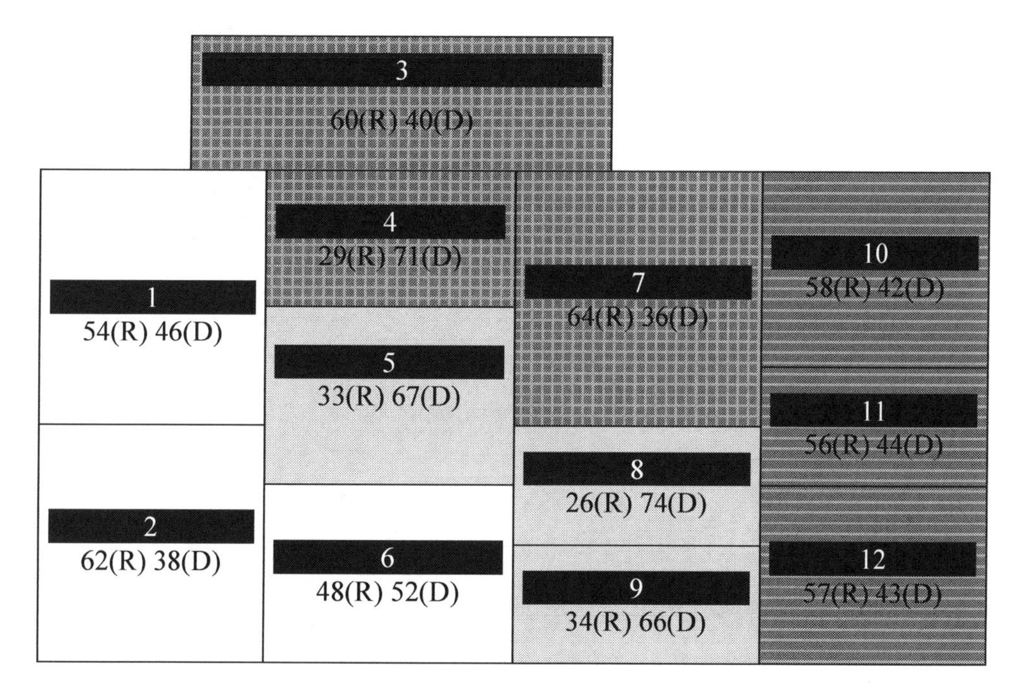

Figure 4: Republican Governor - plan R

District Location	Zone	Republican (000s)	Democrats (000s)	District Location	Zone	Republican (000s)	Democrats (000s)
Top Center	3	60	40	Right Most	10	58	42
	4	29	71		11	56	44
	7	64	36		12	57	43
Lower Center	5	33	67	Left Most	1	54	46
	8	26	74		2	62	38
	9	34	66		6	52	48

Table 5: Republican Governor - plan R

District Location	Zone	Republican Totals and %	Democrats Totals and %	District Location	Zone	Republican Totals and %	Democrats Totals and %
Top Center	3			Right Most	10		
	4				11		
	7				12		
Lower Center	5	93,000	207,000	Left Most	1		
	8				2		
	9	31%	69%		6		

Table 6: Republican Governor - plan R - totals and percentages

5. Calculate totals and percentages for each district and record them in Table 6.

6. Which party would likely win the left-most district (Zones 1, 2 and 6)? What percentage of the votes would the winner probably have?

7. Which party would likely win the right-most district (Zones 10, 11 and 12)? What percentage of the votes would the winner probably have?

8. Which party would likely win the top center district (Zones 3, 4 and 7)? What percentage of the votes would the winner probably have?

A safe district is a district in which the Republican or Democrat will likely win more than 55% of the vote. A very safe district is a district in which the Republican or Democrat will likely win more than 60% of the vote.

9. Which of the districts would be safe for Democrats? Which of the districts would be very safe for Democrats?

10. Which of the districts would be safe for Republicans? Which of the districts would be very safe for Republicans?

Independent Commission Proposal - I

Governor Gerry Manderbilt and state legislature could not reach an agreement. An independent commission was set up to draw the district boundaries. They were asked to construct boundaries so that two Democrats and two Republicans would likely be elected if voting patterns did not change.

11. Form two districts with Republican majorities and two districts with Democratic majorities. The zones in each district must be connected. For example, a district made up of zones 9, 2, and 11 would not be allowed. Shade in Figure 5 as you create your districts. Then complete Table 7.

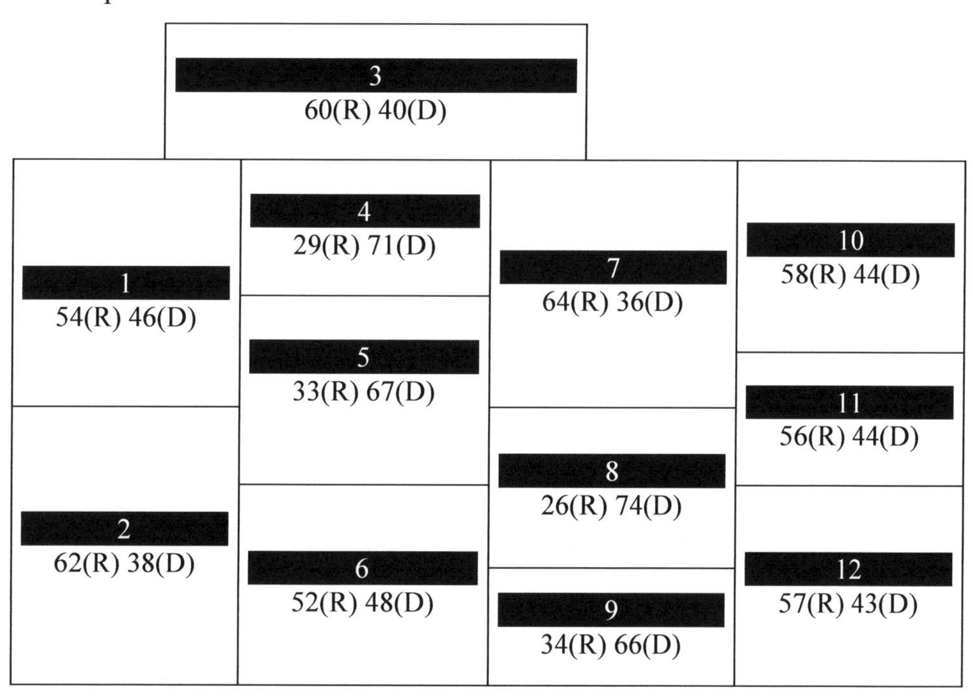

Figure 5: Independent commission - plan I

District Location	Zone	Republican (000s)	Democrats (000s)	District Location	Zone	Republican (000s)	Democrats (000s)

Table 7: Independent commission - plan I

12. Which party would likely win each district? What would the winning percentage probably be?

13. Which of the districts would be safe for Democrats? Which of the districts would be very safe for Democrats?

14. Which of the districts would be safe for Republicans? Which of the districts would be very safe for Republicans?

Project Idea:

Ask students to look at recent elections in their congressional district and determine whether it is considered a safe district. Discuss how the most recent census was used to draw up current congressional districts. What process was used?

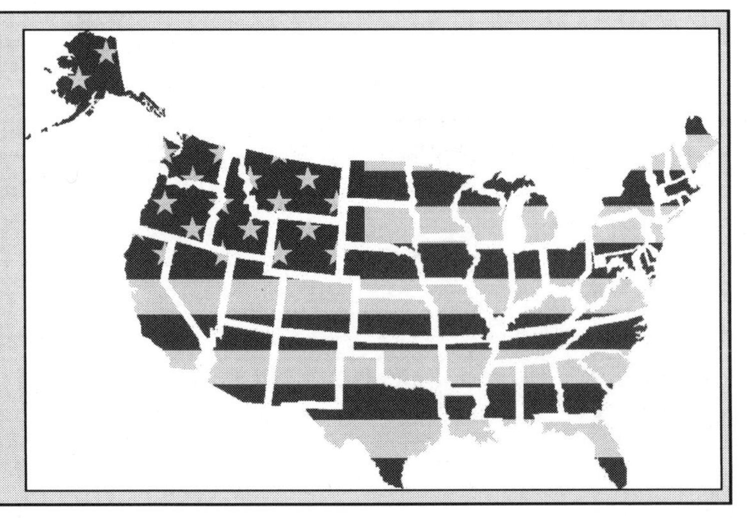

Practice problems

The city of Incognito, CA has a population 64,000. The city is thinking of changing its system of government. Currently, the city is run by a city council of five members. All members are selected in a citywide election. In an election, citizens vote for five people on the long list of candidates. There are sometimes as many as 15 people running for the five seats. The five individuals who receive the most votes become the city council. The individual with the most votes is President of the city council. Because of the large number of candidates, it can be difficult for a voter to learn enough about each candidate to vote intelligently. In addition, because the vote is citywide, some neighborhoods feel they do not have a voice in the city council.

The city is redesigning its government. They plan to have four city council districts and a mayor. Each district will select just one individual to represent it. The mayor will be elected by the entire city. The city is made up of eight neighborhoods as shown in Figure 6. Each neighborhood has an approximate population of 8,000. Approximately 50% of the population voted in a recent US Senate election. The numbers who voted for the Democrat and Republican candidates are recorded in each box. To simplify the analysis, these numbers were rounded to the nearest hundred.

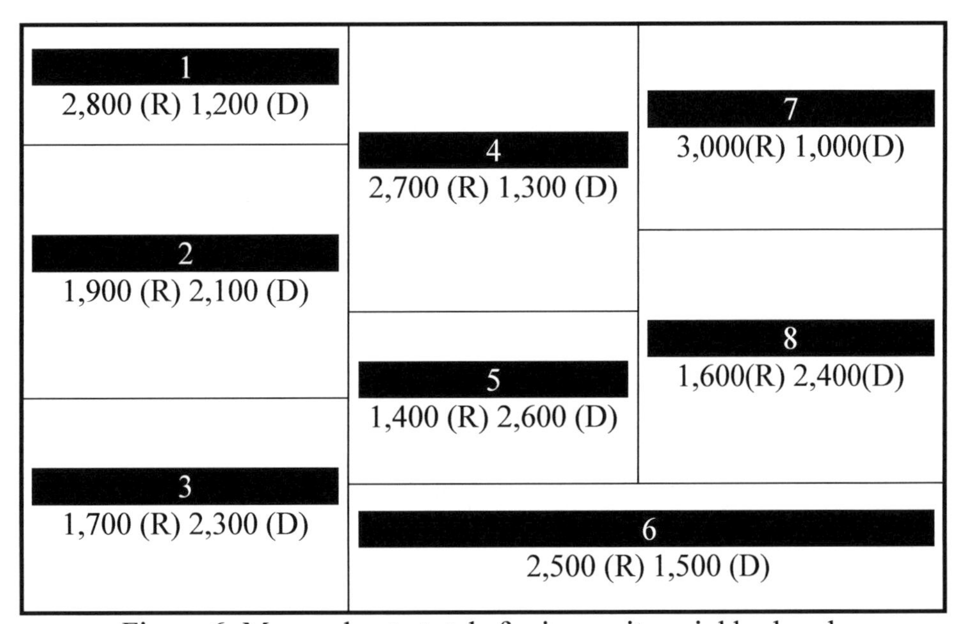

Figure 6: Map and vote totals for incognito neighborhoods

1. What neighborhood had the highest percentage of Republican voters? What was the percentage?

2. What neighborhood had the highest percentage of Democratic voters? What was the percentage?

3. What was the overall percentage of Republican and Democratic voters in Incognito?

Stan Kwo, Renee Hernandez, and Jeremiah Johnson were selected as an independent commission to create the four districts. Each district must consist of two adjacent neighborhoods. After looking at the map, Stan Kwo had a simple suggestion. Combine 1 and 2, 4 and 5, 7 and 8 and finally 3 and 6.

	District			
	A	B	C	D
Stan Kwo	**1 and 2**	**4 and 5**	**7 and 8**	**3 and 6**
Percent Republican				
Likely Winner				

Table 8: Stan Kwo's proposed districts

Both Renee and Jeremiah thought this was a bad idea. They felt it could produce a city council that was biased towards the Republican Party.

4. Complete Table 8 by determining the percentage of Republicans in each district.

5. How many districts would likely elect a Republican city council member?

A very safe district is a district in which the Republican or Democrat will likely win more than 60% of the vote. Renee and Jeremiah decided to see if they could create at least one safe district for each party.

6. Construct a very safe district for Republicans.

7. Construct a very safe district for Democrats.

8. Can you create two more districts with the remaining four neighborhoods? Who would likely win each district?

9. Develop a plan that will likely result in the election of two Democrats and two Republicans.

Activity 15: Teachers' guide

Thinking Through a Lesson Protocol

Standards:

6.RP.A.3.C: Find a percent of a quantity as a rate per 100 (e.g., 30% of a quantity means 30/100 times the quantity); solve problems involving finding the whole, given a part and the percent.

7.RP.A.3: Use proportional relationships to solve multistep ratio and percent problems.

Mathematical Practices:

MP1: Make sense of problems and persevere in solving them.

MP2: Reason abstractly and quantitatively.

MP3: Construct viable arguments and critique the reasoning of others.

MP4: Model with mathematics.

MP6: Attend to precision.

Setting up the problem - Launch	
Selecting tasks/goal setting	(10 minutes) Have students discuss what their congressional district is and who their congressman is. (Look at state map by congressional district if possible). Have students read individually the scenario up through Figure 1.
Questions	What is the population in their congressional district? If data is available from the last presidential election, by congressional district, that might be interesting as well. If not, county data may be available.

Monitoring student work - Explore		
Strategies and misconceptions - Anticipating	**Who - Selecting and sequencing**	**Questions and statements - Monitoring**
(15 Minutes) With Figure 2 and Table 1 displayed side by side, ask students what they notice about the two?		
Complete Table 1.		
Continue reading the text following Imaginary Metropolitan Area.		
(15 Minutes) With Figure 3 and Table 2 displayed side by side, continue reading the Democratic Legislature Proposal-D.		
With Figure 4 and Table 3 displayed side by side, continue reading the Republican Governor Proposal-R.		
Answer questions 1 through 5.		Share out solutions to these answers.
Have students work independently or in pairs to read the Independent Commission Proposal-I. Using Figure 5 and Table 4 answer 6 through 9.		

Monitoring individual student work - Explore		
Strategies and misconceptions - Anticipating	**Who - Selecting and sequencing**	**Questions and statements - Monitoring**
For off-task students or for students that seem to be self-conscious about you listening to them share.		I am just listening or looking to find out how you are working on the problem. This helps me think about what we will do later.
For students that appear to be stuck. Also for when you are having a difficult time understanding their strategies.		Can you tell me a little about your reading? How would you describe the problem in your own words? What facts do you have? Could you try it with simpler numbers?
For students that want to ask you questions, these are ways to uncover their thinking and judge to what extent you want to respond.		Why are you interested in more information about that? Let me say a little about that part. Tell me what you've thought about so far. What do you know?

Managing the discussion - Summarize	
Parts of discussion - Connecting	**Questions and statements - Connecting**
Launching the discussion: Select the problems in questions #1-9 that students are struggling with or you wish to share out.	Will team 1 start us off by sharing one way of working on this problem? Please raise your hand when you are ready to share your solution. What did you do first when you were working on this problem? Let's start by clearing up a few things about the problem. Let's list some key parts in this problem. What was unclear in the problem?
Eliciting and uncovering student strategies	Joe would you be willing to start us off? What have you found so far? Can you repeat that? Can you explain how you got that answer? How do you know? Walk us through your steps. Where did you begin? Can you show us?
Focusing on mathematical ideas	Can you explain why this is true? Does this method always work? How is Bob's method similar to Kelly's method? What do all the solutions have in common? What would happen if I changed the numbers to _____?
Encouraging interactions	Do you agree or disagree with Kahlil's idea? What do others think? Would someone be willing to repeat what Tom just said? Would anyone be willing to add on to what Sue just said?
Concluding the discussion	Can anyone tell me some of the big ideas that we learned today? How would you explain what we learned today to a 5th grader? Some of the key points from our discussion today are . . . Tomorrow we will continue our exploration of ______ beginning with the idea from today that __________.
Post lesson notes	You may wish to assign the practice problems that you feel would benefit the students.

Solutions to text questions

<table>
<tr><th>District Location</th><th>Zone</th><th>Republican
Totals and %</th><th>Democrats
Totals and %</th><th>District Location</th><th>Zone</th><th>Republican
Totals and %</th><th>Democrats
Totals and %</th></tr>
<tr><td rowspan="3">Upper Left</td><td>1</td><td rowspan="3">143,000
47.7%</td><td rowspan="3">157,000
52.3%</td><td rowspan="3">Upper Right</td><td>7</td><td rowspan="3">148,000
49.3%</td><td rowspan="3">152,000
50.7%</td></tr>
<tr><td>3</td><td>8</td></tr>
<tr><td>4</td><td>10</td></tr>
<tr><td rowspan="3">Lower Left</td><td>2</td><td rowspan="3">147,000
49%</td><td rowspan="3">153,000
51%</td><td rowspan="3">Lower Right</td><td>9</td><td rowspan="3">147,000
49%</td><td rowspan="3">153,000
51%</td></tr>
<tr><td>5</td><td>11</td></tr>
<tr><td>6</td><td>12</td></tr>
</table>

Table 4: Democratic Legislature - plan D – totals by district

1. Determine the totals and percentages for each of the two districts in the lower part of the map. Record these values in Table 4. *See table above.*

2. Which party would likely win each of these districts? *Democrats*

3. Under this plan, how many congressional districts would likely be won by Democratic candidates? *Democrats would win all four districts.*

4. Which district would likely have the closest election? *The upper right district. The Democrats would have 50.7% of the vote.*

District Location	Zone	Republican Totals and %	Democrats Totals and %	District Location	Zone	Republican Totals and %	Democrats Totals and %
Top Center	3 4 7	153,000 51%	147,000 49%	Right Most	10 11 12	171,000 57%	129,000 43%
Lower Center	5 8 9	93,000 31%	207,000 69%	Left Most	1 2 6	168,000 56%	132,000 44%

Table 6: Republican Governor - plan R - totals and percentages

5. Calculate totals and percentages for each district and record them in Table 6. *See Table 6.*

6. Which party would probably win the left-most district? What percentage of the votes would the winner probably have?

   The left-most district is comprised of zones 1, 2, and 6. The Republicans would receive 168,000 out of 300,000 possible votes for a 56% majority.

7. Which party would probably win the right-most district? What percentage of the votes would the winner probably have? The right-most district is comprised of zones 10, 11, and 12.

   The Republicans would receive 171,000 out of 300,000 possible votes for a 57% majority.

8. Which party would probably win the top center district? What percentage of the votes would the winner probably have?

   The center district is comprised of zones 3, 4, and 7. The Republicans would receive 153,000 out of 300,000 possible votes for a 51% majority.

9. Which of the districts would be safe for Democrats? Which of the districts would be very safe for Democrats?

   The SAFE district for Democrats would be the one comprised of zones 5, 8, and 9. This zone has 207,000 of the 300,000 or 69%. It is not only SAFE it is VERY SAFE.

10. Which of the districts would be safe for Republicans? Which of the districts would be very safe for Republicans?

The SAFE districts for the Republicans would be the one comprised of zones 1, 2, and 6 with 168,000 out of 300,000 or 56% and the one comprised of zones 10, 11, and 12 with 171,000 out of 300,000 or 57%. None of these are considered VERY SAFE.

11. Form two districts with Republican majorities and two districts with Democratic majorities. The zones in each district must be connected. For example, a district made up of zones 9, 2, and 11 would not be allowed. Shade in Figure 5 as you create your districts. Also complete Table 5.

Answers will vary. One possibility is shown below.

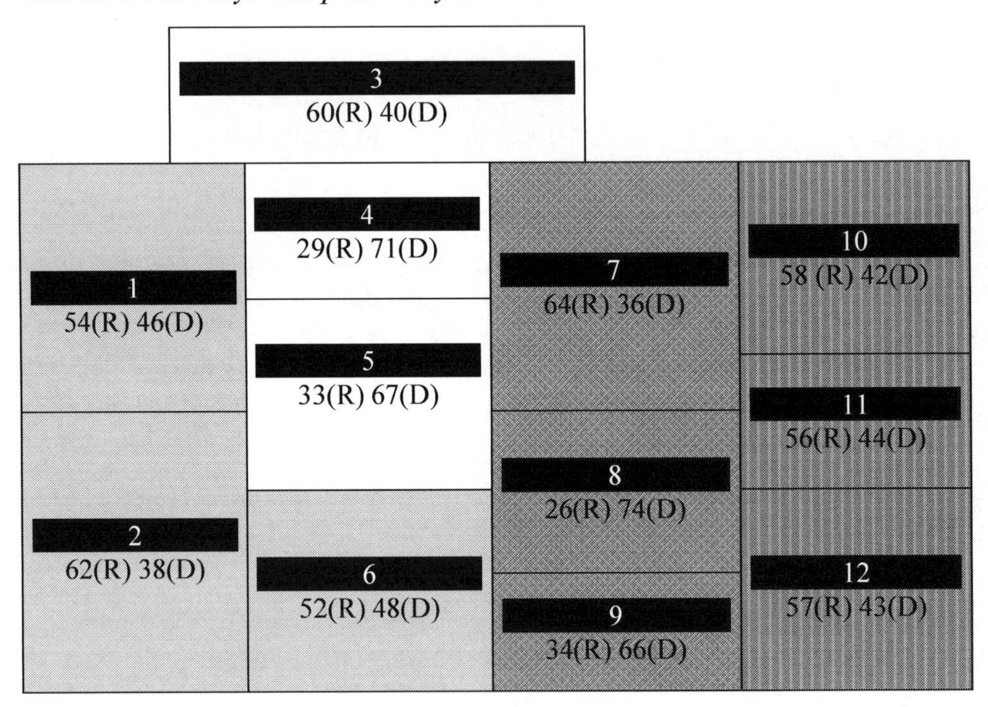

Figure 5: Independent commission - plan I

Answers may vary. One possible scenario would be to combine zones 1, 2, and 6 with 168,000 Republicans and 320,000 Democrats. Also combining zones 3, 4, and5, with 122,000 Republicans and 178,000 Democrats. Combining zones 7, 8, and 9 gives Republican 124,000 and Democrats 176,000 and combining zones 10, 11, and 12 gives 171,000 Republican and 129,000 Democrats.

12. Which party would probably win each district? What would the winning percentage probably be?

Based on the example given in number four, the democrats would win combined zones 3, 4, and 5 with 178,000 out of 300,000 votes or 59%, and combined zones 7, 8, and 9 with 176,000 out of 300,000 votes or 58%. Republicans would win combined zones 1, 2, and 6 with 168,000 out of 300,000 votes or 56% and combined zones 10, 11, and 12, with 171,000 out of 300,000 votes or 57%.

13. Which of the districts would be safe for Democrats? Which of the districts would be very safe for Democrats?

    Both districts would be safe for Democrats. No district would be very safe.

14. Which of the districts would be safe for Republicans? Which of the districts would be very safe for Republicans?

    Both districts would be safe for Republicans. No district would be very safe.

Solutions to Practice Problems

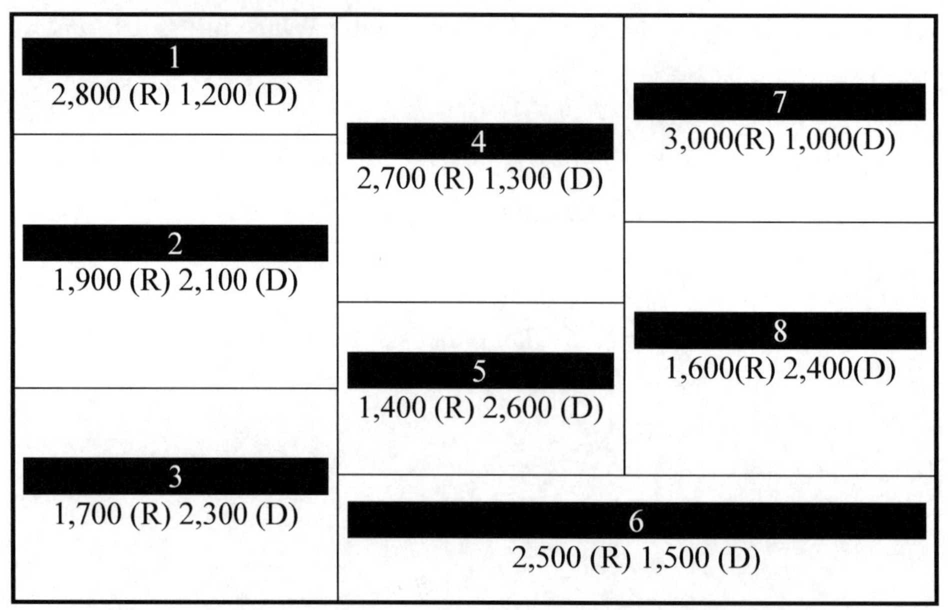

Figure 6: Map and vote totals for incognito neighborhoods

1. What neighborhood had the highest percentage of Republican voters? What was the percentage?

   Neighborhood 7 with (3000 / 4000) × 100% = 75%

2. What neighborhood had the highest percentage of Democratic voters? What was the percentage?

   Neighborhood 5 with (2600/4000) × 100% = 65%

3. What was the overall percentage of Republican and Democratic voters in Incognito?

   Republicans won a total of 17,600 votes out of 32,000. This represents 55% of the total.

	District			
	A	**B**	**C**	**D**
Stan Kwo	**1 and 2**	**4 and 5**	**7 and 8**	**3 and 6**
Percent Republican	*4700 / 8000* *58.8%*	*4100 / 8000* *51.3%*	*4600 / 8000* *57.5%*	*4200 / 8000* *52.5%*
Likely Winner	*Republican*	*Republican*	*Republican*	*Republican*

Table 8: Stan Kwo's proposed districts

4. Complete Table 8 by determining the percentage of Republicans in each district.

   See Table.

5. How many districts would likely elect a Republican city council member?

   Four.

6. Construct a very safe district for Republicans.

   Answers will vary. Combine neighborhoods 4 and 7. 5700 / 800 which is 71.3%.

7. Construct a very safe district for Democrats.

   Combine neighborhoods 5 and 8. 5000/8000 which is 62.5%.

8. Can you create two more districts with the remaining four neighborhoods? Who would likely win each district?

   If the answer to question 5 was a district combining neighborhood 1 and 4 and then formed a very safe Democratic district, it would not be possible to form two districts of adjacent neighborhoods. With the answer above the remaining two districts would combine 1 and 2 and then 3 and 6. Both would have Republican majorities.

9. Develop a plan that will likely result in the election of two Democrats and two Republicans.

   Combine 1 and 4 – likely Republican, 7 and 8 likely Republican, 2 and 3 likely Democrat, 5 and 6 likely Democrat.

Made in the USA
San Bernardino, CA
19 March 2018